T0383178

A Research Agenda for Construction Management

Elgar Research Agendas outline the future of research in a given area. Leading scholars are given the space to explore their subject in provocative ways, and map out the potential directions of travel. They are relevant but also visionary.

Forward-looking and innovative, Elgar Research Agendas are an essential resource for PhD students, scholars and anybody who wants to be at the forefront of research.

For a full list of Edward Elgar published titles, including the titles in this series, visit our website at www.e-elgar.com

A Research Agenda for Construction Management

Edited by

ROINE LEIRINGER

Associate Professor, Department of Real Estate and Construction, The University of Hong Kong, Hong Kong

ANDY DAINTY

Pro-Vice Chancellor for Education, Manchester Metropolitan University, UK

Elgar Research Agendas

Cheltenham, UK • Northampton, MA, USA

© The Editors and Contributors Severally 2023

All rights reserved. No part of this publication may be reproduced, stored in a retrieval system or transmitted in any form or by any means, electronic, mechanical or photocopying, recording, or otherwise without the prior permission of the publisher.

Published by
Edward Elgar Publishing Limited
The Lypiatts
15 Lansdown Road
Cheltenham
Glos GL50 2JA
UK

Edward Elgar Publishing, Inc.
William Pratt House
9 Dewey Court
Northampton
Massachusetts 01060
USA

A catalogue record for this book
is available from the British Library

Library of Congress Control Number: 2023936969

This book is available electronically in the **Elgar**online
Geography, Planning and Tourism subject collection
http://dx.doi.org/10.4337/9781800375451

Printed on elemental chlorine free (ECF)
recycled paper containing 30% Post-Consumer Waste

ISBN 978 1 80037 544 4 (cased)
ISBN 978 1 80037 545 1 (eBook)

Printed and bound in the USA

Contents

Figures

Tables

Contributors

Dominic D. Ahiaga-Dagbui, Senior Lecturer in Construction Management, School of Architecture and Built Environment, Deakin University, Australia.

Mike Bresnen, Professor of Organisation Studies, Department of People and Performance, Manchester Metropolitan University, UK.

Paul W. Chan, Professor of Design and Construction Management, Department of Management in the Built Environment, TU Delft, the Netherlands.

Vivien Chow, Lecturer in Construction and Project Management, School of Architecture, Building and Civil Engineering, Loughborough University, UK.

Andy Dainty, Pro-Vice Chancellor for Education, Manchester Metropolitan University, UK.

Dilek Ulutas Duman, Senior Lecturer, Department of Architecture and Civil Engineering, Chalmers University of Technology, Sweden.

Stuart D. Green, Professor of Construction Management, School of Construction Management and Engineering, University of Reading, UK.

Stephen Gruneberg, Honorary Professor, Bartlett School of Sustainable Construction, University College London, UK.

Eyüphan Koç, Assistant Professor of Civil Engineering, Boğaziçi University, Turkey.

Mathangi Krishnamurthy, Associate Professor of Anthropology, Department of Humanities and Social Sciences, Indian Institute of Technology Madras, India.

Roine Leiringer, Associate Professor, Department of Real Estate and Construction, The University of Hong Kong, Hong Kong.

Ashwin Mahalingam, Professor of Civil Engineering, Department of Civil Engineering, Indian Institute of Technology Madras, India.

Igor Martek, Senior Lecturer in Construction, School of Architecture and Built Environment, Deakin University, Australia.

Eunice Maytorena-Sanchez, Senior Lecturer in Project Management, Alliance Business School, University of Manchester, UK.

Johan Ninan, Assistant Professor, Faculty of Civil Engineering and Geosciences, TU Delft, the Netherlands.

Evangelos Pantazis, Computational Design Lead, IBI Group and Topotheque Design Research, USA.

Steve Rowlinson, Professor Emeritus of Construction Project Management, Department of Real Estate and Construction, The University of Hong Kong, Hong Kong.

Daniel Sage, Reader in Organisation Studies, School of Business and Economics, Loughborough University, UK.

Libby Schweber, Professor of Sustainable Construction, School of Construction Management and Engineering, University of Reading, UK.

Natalya Sergeeva, Associate Professor in the Management of Projects, Bartlett School of Sustainable Construction, University College London, UK.

Lucio Soibelman, Professor of Civil and Environmental Engineering and Spatial Sciences, University of Southern California, USA.

Graham M. Winch, Professor of Project Management, Alliance Business School, University of Manchester, UK.

1 Construction management research: a community at a crossroads?

Roine Leiringer and Andy Dainty

Introduction: a personal conviction

Setting out a research agenda for an eclectic community comprising several thousand academics is a daunting task. We must admit that we thought long and hard about it before we accepted the invitation from the publishers. The idea that we would be better placed to ascribe such an agenda than any number of other academics in this applied area of research challenged our shared view of the field. We see construction management research (CMR) as a broad church of researchers from a multiplicity of backgrounds and orientations, a plurality that should be celebrated rather than lamented. But despite this multiplicity, there remains a need for our community to better define the ways in which its contributions to theory and practice are made in order to ensure that knowledge accumulates and the impact of our work does justice to the access that we are so often afforded to the institutions, organisations, and projects that constitute the sector. We set out this argument back in 2017 (Leiringer and Dainty, 2017) where we made a clarion call for the community to set out a new research agenda to address this failing. Five years on, we have invited a diverse array of authors, whose work we feel reflects the kinds of scholarship we were advocating for, to refine these arguments and to set out a more granulated articulation of what is needed. We see their contributions setting out fresh directions for our community that will hopefully resonate with, and build upon, the myriad research positions and paradigms that define our eclecticism, but in ways that ensure the impact of our work in relation to both theory and practice.

We are both longstanding members of the CMR community and passionate advocates of what it contributes to both the theory and practice of the creation of the built environment. We have over the past two decades been active scholars in terms of driving our own research endeavours, as well as supervising

PhD students and post-doctoral researchers. We have both been privileged in terms of being able to participate in a host of academic activities, such as examinations, exchanges, and conferences, as well as taking part in national and international research initiatives. Through these activities, we have made many friends and established meaningful and sustained research collaborations. Coming from different academic backgrounds ourselves, we have both sought to draw upon theoretical ideas from a number of fields in shaping our own research practice, and have also encouraged others to do so. Perhaps most notably, we have had a longstanding association with the journal *Construction Management and Economics*, first as editors for about ten years, and then as editors-in-chief from 2017 to 2019. All this notwithstanding, and although we have worked together since being involved with a major research project back in 2004, we certainly do not always agree with each other, and we like to challenge each other's perspectives on regular occasions. We are, however, united in our belief that our community could – and should – do better when it comes to defining and furthering theoretical debates and influencing the broader research communities from which we draw. Whereas once the field could be considered nascent and in flux, in our view, it is now established and large enough to begin to define clear, theoretically informed debates around which communities of research practice and accumulations of knowledge form; and yet these are often difficult to identify within the amorphous array of heterogeneous activities that characterise CMR.

A point of departure

Our point of departure for this book is a shared view that our research community is at a crossroads. CMR academics often reside in faculties where research funding is an expectation, but where opportunities to draw down such funding are scant. At the same time, many of the field's journals are not seen to be of sufficiently high quality to attract the very best papers and, by expansion, are not routinely cited within the journals from whose insights they draw. This is not to say that the CMR community suffers from a lack of engagement with its work, as we enjoy a level of industry connectivity and collaboration that is much stronger than in many other fields. Indeed, many academics benefit from strong knowledge transfer funding and consultancy ties that double as opportunities for conducting research. However, this results in a tendency for researchers to prescribe solutions *to*, rather than seek understanding *of*, the problems that they attend to. As one of our industry collaborators said to us years ago, "Don't tell me how to run my company, tell me the difficult questions that I should be asking myself!" This is something that we have taken to heart, and we would never engage with senior industrialists by attempting to tell them how to do whatever it is that they do. To the contrary, we as academics must understand, and appreciate, what advice we can offer on the basis of the research that we have conducted and, in prolongation, how we could best contribute to the advancement of construction practice.

There is a fine line, we believe, between offering opinions and academic insights. If we cannot separate the two, then there is nothing separating us as academics from us as consultants. We would argue that it, at times, is enough to simply 'hold up a mirror' and challenge practice, drawing upon theoretical insights to inform this analysis. Where we do suggest interventions for shaping practice, these must be founded on sound empirical research as well as a strong theoretical underpinning and, in turn, contribute to the further development of that theory in ways that can be abstracted from the site of intervention. This is a contribution often lacking in many papers published within CMR journals, which serves to reinforce our standing as a poor relation in the discipline hierarchy. Challenging this status quo is what this book sets out to achieve.

It is a truism that the ways in which construction management academics interact, such as through their publications, reviews, and supervisions, lead to the reproduction of methods and methodologies, and agreement around theories. Some would see this as a demonstrable sign that the field is maturing. But there is also an inherent danger in such an outlook – that we fail to inform our work with ideas and approaches that might be more developed than our own, and that we do not subject these theories to challenge through our unique empirical context. Indeed, the tendency within CMR is to reproduce the same approaches that have served the community well in the past, but that arguably hold back the field's evolution. We could point, for example, to the plethora of papers developing and espousing so-called 'Critical Success Factors', tools for informing what to focus on to ensure success in all aspects of construction. Some might argue that such work, if well executed, plays an important role in shedding light on the heuristics that inform construction, effectively abbreviating our understanding of known practice. But while this might have internal value to our community, the extent to which this has any resonance outside of our field is questionable. There seems, therefore, to be a tendency to look inwards to replicate and refine approaches that fail to challenge, or theoretically inform, the norms of practice and the alternative approaches from which researchers might draw. Dustbowl empiricism might thrive under such a paradigm, but generalisable and abstractable theoretical thinking will not. We can contrast this kind of work with the kinds of scholarship advocated by Van de Ven (2005) whereby the process of iterating and fitting between problem formulation, theory building, research design, and problem solving is predicated on engaging with others throughout the research process, whether that be other experts or users of the research.

We would argue that breaking out of the replication trap and seeking insights from outside of the CMR field is a necessary transition and one which reflects a deepening maturity of our research community. This is a theme that we have written about extensively in our *Construction Management and Economics* editorials, and which has underpinned the approaches of those who have contributed to this volume; all challenge us to look outside of our domain to redefine the agendas and debates that will evolve the field. Some are long-established researchers within our field whom we have long respected and 'grown up' reading. Others are more nascent scholars whose work we have come to know more recently and who have begun to question some of the historically dominant nostrums through their own research practice. A third group comprise researchers from what might be considered adjacent or related research fields. They come from communities that draw upon different framings in engaging with the construction sector as an empirical or theoretical arena and, thus, hold different perspectives on construction to those of us who have been heavily involved in, and conditioned to, CMR for an extended period. The book therefore lays out a wide range of perspectives that draw upon diverse theoretical and empirical insights. What unites all of them, however, is that they provide more than just a critique of how dominant approaches limit our understanding; they provide alternative views and set out a clear research agenda as to how we might collectively address the failings outlined above. This is not to tell the reader *how* to do research, but to signpost them to the avenues that might define new, more prosperous opportunities for our community.

Conceptualising construction management research

Construction shares many features present in other sectors, but the ways in which they come together renders it unique. Firms work together as part of a complex ecosystem to erect, repair, and demolish all types of building and civil engineering structures and, with few exceptions, activity takes place at the point of purchase. This contrasts with manufacturing where goods tend to be fabricated before being moved to a sales location and then sold to the customer. The so-called 'engineer-to-order' supply chains that emerge from this unusual delivery model render the industry very different from many others (Gosling et al., 2014). Complex constellations of heterogeneous organisations of all shapes and sizes come together as part of a temporary delivery system before disbanding and moving on to other endeavours. At any one time, firms might be involved in numerous projects, spread both by project type and geography, and at different stages of development.

A core characteristic of the construction sector is that it comprises many firms operating in distinct markets, thereby forming a network of interrelated markets for equipment hire, labour, materials, and components. The temporal nature of these supply networks is writ large once again, but here as the locus of intersecting products and components that together make the completed product. Given this complex backdrop, the procurement of work and contractual transactions are similarly complex. Often based on an auction system where bidders compete for the work on price and performance, the high degree of risk involved in winning work has led to a reliance on subcontracting and the industry being dominated by very small firms. It is noteworthy, however, that the relatively few large firms still contribute a comparatively large part of the construction value output. In the UK, for example, 0.1 per cent of the firms account for around 25 per cent of the total turnover in the industry (Gruneberg and Francis, 2019).

Construction, in many ways, epitomises a project-based industry. Decentralisation and dispersed modes of working in inter-organisational projects are defining characteristics, with procurement commonly focusing on price over value, a situation compounded by its clients often not necessarily being the end-users. Projects are often highly customised, rarely undertaken within a standard framework, and are characterised by long supply chains, extensive use of subcontracting, and an emphasis on short-termism (Dainty et al., 2017). The phasing of project activity, and the transient and time-pressured nature of project work, leads to regular secondment and movement of staff between projects being common. Hence, production involves not only non-routine production processes, but also complex inter-professional and inter-organisational contractual and working relationships that govern project-based interaction. This in turn has led to the formation of deeply ingrained institutional practices and norms that heavily influence the management of construction projects (Leiringer et al., 2022).

It is clear from the preceding discussion that providing a straightforward definition of what is and what is not included in the 'construction sector' is anything but trivial. Construction activities involve combinations of actors, such as clients, architects and designers, contractors, engineering and specialist consultants, material and plant suppliers, and invariably actors such as governments, professional organisations, and the real estate, finance, and insurance industries, and more. While, from an economics perspective, it is possible to provide a definition on the basis of statistical classifications (see Gruneberg,

Chapter 2, this volume), we would agree with Green (2011) that the boundary that is drawn around the 'construction industry' is something that is continuously renegotiated. This leads us to the question of how we could possibly confine the term 'construction management research'. Quite a number of years ago now we, together with Will Hughes, tried to do this for *Construction Management and Economics* when we were tasked to come up with a new aim and scope for the journal. We tried numerous different versions before we settled on:

> Our concern is the production of the built environment. We seek to extend the concept of construction beyond on-site production to include a wide range of value-adding activities and involving coalitions of multiple actors, including clients and users, that evolve over time. We embrace the entire range of construction services provided by the architecture/engineering/construction sector, including design, procurement and through-life management.

With hindsight, we accept that, in trying to be inclusive and all-encompassing, this 'definition' lacks precision. Incorporating so much leaves the boundaries of the field weakly defined and subject to multiple interpretations and reconceptualisation. The industry being so hard to pin down might be one reason why researchers within our community often focus their work on big, powerful actors, rather than those smaller firms that form the rump of the sector. Arguably, focusing on these large firms (incumbents) helps cement a status quo in the construction industry where minor changes in practice might be frequent, but where the structure and power balance remains the same over time (see Leiringer, 2020).

Defining the field: CMR today

Several attempts have been made to provide a history of construction management, both in terms of teaching and research. We have, over time, read many of these, and we have also been fortunate to have met many of the people who have been singled out as particularly significant pioneers in establishing both research and teaching programmes. All the stories notwithstanding, one thing we can say with confidence is that there is no definite linear development of CMR, but many loosely connected developments in areas such as building economics, construction engineering, and operations management. These started to coalesce in the 1980s under the common English term of construction management (e.g. Langford and Hughes, 2009; Harty and Leiringer, 2017). Ever since, there has been significant growth in the field, with a proliferation of specialist conferences and interest groups. For example, organisations such as the International Council for Research and Innovation in Building and Construction (CIB) w55/65, the Association of Researchers in Construction Management (ARCOM), the Australasian Universities Building Education Association (AUBIA), Construction Researchers on Economics and Organisation (CREON), West Africa Built Environment Research (WABER), and online networks such as the Co-operative Network of Building Researchers (CNBR) provide for thousands of academics to evolve the field. Construction management is also a popularly taught programme at both undergraduate and postgraduate levels. It could be argued that the unity in activities and strength of the community was at its strongest in the late 1990s and early 2000s. During this period there was significant research activity in many countries across several continents, and international conferences attracted large numbers of participants. However, during this time the community also expanded and, over the past 20 years, it has been diversified. CMR is certainly no longer merely an extension of operations management research in the construction context.

What construction management academics focus on today is probably most apparent as a reflection of the themes that have dominated CMR conferences in recent years. For example, Table 1.1 shows the themes that drove the 2008 ARCOM conference against those from 2021. With the exceptions of building information modelling (which is in itself a subset of information technology (IT) and support systems), disaster management and resilience, and offsite construction, there has been virtually no change in 13 years. That similar research themes have endured says as much about the CMR field as it does the industry itself.

In their content analysis of some 3500 construction management publications between 2000 and 2020, Bilge and Yaman (2022) show that the most popular topics are: (1) building information modelling; (2) information management; (3) scheduling optimisation; (4) Lean construction; (5) cost optimisation; (6) agile approach; and (7) megaprojects research. Thus, the field has broadly developed along a number of sub-strands. The first focuses on the organisation and management of construction as a productive activity. This borders a range of subfields from organisation theory through to operations and human resource management. This focus on the social side of construction endeavours draws upon a plethora of theoretical ideas, from hard production models to softer relational and cultural constructs. A second strand focuses on the development of materials and technologies necessary to bring about a transformation of the construction process. From novel forms of production through to digital practices, this technical oeuvre of research focuses on systems and products which can help transform construction as a process and

Table 1.1 Themes of the ARCOM Conference, 2008 and 2021

ARCOM Themes 2008	ARCOM Themes 2021
• Inter-organisational relations and supply	• Building information modelling
• Human behaviour and culture	• Equality and diversity
• Business performance	• Human resources management
• Cost and financial management	• Information management
• Knowledge management	• Infrastructure development
• Human resources and skills	• Offsite construction
• Equality and diversity issues	• Planning, productivity, and quality
• Risk	• Research and education
• Organisation strategy and management	• Sustainability in the built environment
• Project performance	• Construction design & technology
• Education and learning	• Disaster management and resilience
• Sustainability, environment, and green issues	• Health, safety, and well-being
• Project management	• Law and contracts
• Economics and industry	
• Information technology and support systems	
• Health, safety, and respect for people	
• Research methods	
• Procurement	
• Law and contracts	

a product. Sitting between these strands, there is a smaller but nevertheless significant body of work that can loosely be described as socio-technical in nature. Focusing on understanding the ways in which people, processes, technologies, and productive activity interact, and how this manifests both within and through practice, this work is less focused on enabling more productive outcomes from a particular stakeholder perspective, and instead emphasises understanding the socio-material relations that define such outcomes.

We do, of course, recognise that crude aggregations of CMR fail to acknowledge or reflect both the nuances of the work contained therein and the significant body of work that sits outside of these definitions. However, they do serve as a framing device against which we might classify the contributions of the field. Today, CMR academics are heterogeneous in terms of background and beliefs on how research should be carried out. Yet, they do, nonetheless, coalesce to form recognisable organisations (e.g. research groups, divisions, departments, and schools). Hence, rather than necessarily being united by common ontological, epistemological, and methodological understandings, CMR academics are joined together by a system of social relations, or through their attention to a particular sectoral context. They regularly interact through multiple activities, such as: the refereeing process; co-authorship of books, conference papers, and journal articles; visiting professorships; the external examination system; membership on various committees and professional bodies; participation in national and international conferences; and numerous seminars, panels, and events. Accordingly, it is possible to conceptualise CMR as an arena with particular co-evolved logics representative of the configuration, coherence, interests, and formation of its members over time. Hence, while there might not be a clear-cut definition of the term 'construction management', CMR can nonetheless be conceptualised as an applied academic field, with a relatively easily defined membership and set of social dynamics.

Rigour and relevance - complex bedfellows

Academic research in applied fields like construction management uses real-world contexts as sites for both developing research questions and conducting empirical studies to examine them. In return for such access, researchers make claims that research outputs will have utility outside of academic domains. Without such promises of exchange, the relationship between research and industry becomes essentially parasitic, which does little to develop good relations between the two. This has prompted a debate within the broader management literature of how to satisfy the so-called double hurdles of being relevant to industry and maintaining academic rigour, or of bridging the 'relevance gap' between the two (Pettigrew, 1997). Extending this debate to the role of the academic in shaping the construction sector is not trivial (see Bresnen, Chapter 3, this volume). To start with, we should bear in mind that the desires of CMR academics to actively seek to influence their surroundings vary. Clearly, they do not share the same ontological and epistemological positions regarding the methods they mobilise and how the products of their research contribute to academic or practitioner knowledge, and why would they? Second, the utilised mode of research will influence and shape the impact of the research outputs. It will also, at least partially, affect the individual researcher's inclination to actively seek higher and wider impact for their work. Finally, we also need to take into consideration how research findings are received by those who are to be affected. Needless to say, few in industry care about the number of citations a research collaboration might generate. Much more important are the direct benefits that might accrue from engaging in research and the perceived competitive advantages that these might bring. There is, as such, little doubt that industry prefers short-term impact ahead of potential for long-term impact. Such sentiments are also apparent amongst the champions of 'knowledge exchange' and those who call for CMR to have a greater impact on industry. Thus, research focusing on solving current industry problems, and hence with supposedly high relevance, is favoured.

Judging relevance from a more critical perspective requires a more involved discussion of the sorts of contexts in which research-generated knowledge must make sense. It goes without saying that research outputs could benefit individuals, single firms, sectors, 'industry' as a whole, and 'the economy' or 'society' at large. Yet, at each of these levels, multiple perspectives co-exist. Research outputs and the beneficiaries thereof should not, therefore, be conceptualised as unidirectional or asymmetrical. It follows that in an applied field, such as construction management, questions can justifiably be raised about who the ultimate beneficiaries of research are supposed to be, and who are denied any advantage gleaned. This is especially pertinent to research-council-funded projects. Funding is provided on the prerequisite of a commitment to enhancing the competitiveness of a national construction industry in general, but often specifies that researchers should work closely with specific commercial organisations. This leads to a host of philosophical and ethical problems, such as: offering an advantage to some firms at the expense of others; and how to balance requirements of confidentiality and wide dissemination of findings and knowledge within both academic and practitioner domains.

In summary, for CMR scholars to influence the construction sector, there is a need to go beyond mere academic accommodation of, and orientation towards, industry needs. In particular, there needs to be clear recognition of the importance of producing academic insights that are relevant, partly because they are not constrained by the immediate pressures of business. Here we would argue that to call for increasing the impact of research is not the same as prioritising relevance ahead of rigour, even if debates tend to polarise the two. There is nothing to say that rigorous research cannot have a significant impact, either in the shorter or longer term. If anything, the processes through which new knowledge claims diffuse through practitioner landscapes are complex, and the potential for and degree of long-term impact is hard to assess. We would argue that following the path of least resistance of complying with external pressures and aligning research solely with short-term business priorities should be resisted, or research becomes akin to consultancy. The shaping and developing of a research agenda for construction management should be grounded in the understanding that, in practice, business and academia reflexively interact and influence each other. Priorities of firms are not always in harmony with wider society, yet CMR cannot escape its connections to both academic and industrial arenas.

Challenging the field

And so, what does our field need to do to address some of the issues described above, and what will it mean for CMR if it does not shift to a new way of being? If we look at the field globally, our community is diffuse, both in terms of a lack of critical mass in geographically proximate areas, and in the density of activity generating a cogent body of theory around specific issues and topics. Thought leadership, while quite clearly present within the field, tends to be limited to domain-specific debates, rather than generalising to theoretical areas that resonate *across* multiple application domains. In other words, the field talks to itself, but strangely not in a way that accumulates and builds upon the knowledge that it generates very well. This lack of coherence and accumulation perpetuates the proliferation of topics, many of which come around again. In short, the field grows and diversifies, but the debates stay largely the same. When you are neither rigorous nor relevant, you clearly have little value, but it also seems that even where the research is rigorous and the debate is relevant, CMR often suffers from a neglect of theory, stopping at the rich description of a single case or couple of cases, with no reflection on the implications of that analysis for more abstract analytical frameworks or general understandings. While some of this can be laid at the feet of a weak engagement with the literature (cf. Schweber, 2015), breaking out of this delimiting tendency will require a different kind of thinking and engagement in the future.

What we see is a field in need of change and re-emphasis. These changes extend from a basic need for reflexivity around the kinds of knowledge that we produce and reproduce through our research practice, to the ways in which we draw upon theory from other fields (see Schweber and Chow, Chapter 4, this volume) to the ways in which we seem to ignore it (Bresnen, Chapter 3, this volume). We also see an acute need for the community to decolonise; many problems are rooted in a UK/European/North American context and show little regard for different socio-economic and socio-cultural contexts. Furthermore, even what might be termed good technical research veers towards software development and consultancy, or even to technology application, rather than as discovery or theory building around that technology (see Ahiaga-Dagbui and Martek, Chapter 10, this volume).

We see CMR as having arrived at an important fork in the road. It could continue its current trajectory, uncritically borrowing from other fields, selectively applying these ideas to the construction industry context, and largely talking to itself. Or it can shift its approaches, build on theory developed elsewhere to join these critical debates as opposed to observe them, and in doing so attract researchers from other domains to consider the construction sector as an interesting empirical and theoretical arena. This latter pathway will demand that CMR researchers pay attention to the origins and definitions of the constructs imported, how they can be operationalised, and how they might, or might not, be compatible. This is not only necessary in terms of making contributions to the extant literature, it is also fundamentally important in achieving the required convergence within the construction management literature for research work to start to be accumulative and have practical purchase outside the immediate organisational context studied.

CMR: new directions

The challenge posed above brings us to our chapter contributions, each of which offers a critique of the dominant orthodoxy in CMR and a challenging new perspective on how we might re-think our research approaches and agendas.

What is it that we are researching?

We open with Stephen Gruneberg who explains the power that an economics perspective should have on contextualising and revealing the production tensions that characterise construction activity, and especially in understanding the relationships between the firms that constitute it. His chapter shows construction as a collection of loosely coupled subsectors, and as an industry constituted by differing business models and competitive behaviours. He sets out an argument that we cannot tackle the issues that the industry faces without first understanding why the industry functions in the way that it does. The imperfect ways in which construction firms compete – the industry's economic model – arguably underpins much of CMR, and yet the economic plight of the firm is poorly understood. More attention to this, if it is based on rigorous methods and approaches, would provide a much sounder basis for the interventions that many argue are required to improve the industry. He sets out the agendas that would benefit from this understanding, from IT to skills, to planning to carbon reduction, each of which must be underpinned by a sound understanding of the economics of the industry and the multiple markets that constitute it. Gruneberg's chapter serves as a useful backdrop to the contributions that follow in that it encourages the reader to consider what the domain is that we seek to understand, and questions some of the taken-for-granted aspects of the sector's operation that underpin so much of the extant body of work. This is a theme that plays out across several of the contributions that follow.

Rigour and relevance

In Chapter 3, Mike Bresnen explores rigour and relevance in CMR, and, specifically, the extent and ways in which the theories developed in business and management research (BMR) are borrowed and applied, and what this means for the production of knowledge in the field. This is important to furthering understanding of construction, and there is much more to do in mobilising insights from adjacent and related fields; but doing this well arguably demands a very different approach to that which dominates the CMR field at present. Bresnen explores the rigour-versus-relevance debate in BMR, as well as the surrounding modes of knowledge production, both of which have shaped a similar, if relatively nascent, debate within CMR. However, he also explains the very different institutional landscapes which characterise the BMR and CMR domains. CMR, he argues, faces an additional hurdle of engaging with a contested knowledge base as well as those of rigour and relevance. Bresnen argues that CMR should embrace hybridisation, whereby the strengths of the field can be retained while engaging with broader BMR debates and critical agendas. But doing this has significant implications at the institutional and relational levels for CMR that would compel it to engage with new networks well beyond its traditional community of practice. Here lies a significant challenge for those of us within the CMR field who have operated in a fairly insular way, but one which we must address if our work is to pervade the fields from which we draw.

Drawing on similar themes as Bresnen, but drilling down to examine some of the methodological implications that flow from such consideration, Schweber and Chow in Chapter 4 explore the role that qualitative enquiry should play in opening up new ways of thinking about construction management. Qualitative work often fails to achieve this because of the ways in which it is undertaken within CMR. Their argument hinges on the assertion that, rather than theorise around their matters of concern, construction management researchers instead tend to draw on their observed empirical insights to suggest what practitioners should do to serve a particular agenda. Greater (and better) use of theory will, they argue, address both issues of rigour and relevance. In other words, CMR should be characterised by epistemological rigour and be informed by theory. Theory development flows from the ongoing comparative processes of exploring its relevance to different contexts, but may also serve the different purposes of explaining, ordering, comprehending, enacting, or provoking. Appropriately used, qualitative research should go beyond the mobilisation of a way of understanding (the realm of the consultant) to offering insights which surprise, or even reconceptualise, the very problems that we study. The challenge to our community is that CMR must move away from the dominant positivist framings of interpretation characterised by representationalism towards a more heuristic use of theory. Schweber and Chow remind us that 'one case does not a theory make'; it is the CMR field as a whole that must address this failing as a collective if we are to theorise phenomena in ways that yield more rigorous, robust, and interesting outcomes.

Challenging dominant orthodoxies

An example of how we might do what Schweber and Chow suggest is provided in the fifth chapter by Stuart D. Green and Dilek Ulutas Duman, who extol the virtues of narrative analyses as a way of questioning sector policy and the reform discourse. They argue that there is a tendency for the CMR field to take policies and the assumptions on which policies are based for granted rather than to understand the power of narratives as the conduit through which policy and practice are enacted. They further argue that construction management researchers often become trapped in needing to follow the dominant policy orthodoxy to attract funding opportunities, which only serves to further reinforce the very narratives that are arguably in need of change. Green and Duman therefore point to the need to understand the formation of the narratives themselves and how this enables insights into the processes of collective sensemaking and, by extension, the ways in which meaning is ascribed to individual and collective action. They illustrate the identification of the narrative 'plot structures' through the example of modern methods of construction and the ways in which this serves a particular reform discourse and, by extension, subjugates others. Their message is that the CMR community has an important role to play in holding recurrent narratives, and especially the hubris of those shaping it, to critical scrutiny. Such a perspective illustrates the role that stronger theorisation and the need to engage with broader critical debates can have within the sector.

Our next set of chapters challenges dominant research objects and abstraction levels in CMR. For many CMR academics, focus is on the delivery of projects, often around ensuring the achievement of performance requirements that serve particular constituencies and situations. In Chapter 6, Paul W. Chan challenges this orthodoxy, broadening the focus to look beyond project boundaries to consider different scalar and temporal possibilities that foreground other concerns, such as those relating to projects as a creative endeavours. Chan's perspective is challenging, not insomuch of the need to consider other production concerns, but rather in setting out projects as innovation rather than simply as arenas for innovation to occur. Further, he challenges the field to consider the myriad interrelationships that exist within the broader constellation of projects and organisations that surround them, setting projects within their broader institutional context. This project ecosystem view has received relatively little attention in CMR and yet opens up new possibilities to reconceptualise projects as sites of experimentation, fluidity, and influence over broader matters of concern once considered to lie well outside of the project boundary. This multi-scalar perspective yields new possibilities to view projects and projectification more broadly as ongoing and open-ended rather than as bounded endeavours. It also requires CMR to stray beyond its own intellectual hinterlands and comfort zones to engage with the fields to which its institutional contexts relate.

In Chapter 7, the theme of project organisation and, specifically, the relationship between cognition and action in complex environments is developed further. Maytorena-Sanchez, Sergeeva, and Winch explore how project actors think about the future. Their point of departure is that projects research has largely progressed along the parallel tracks of coordination, systems, contracts, and decision-making, but that an information processing approach as espoused by the Carnegie School enables us to bridge these perspectives by seeing projects as a problem in information. They bring together thinking around cognition, decision-making, and action in the inherently uncertain and unfolding environment that characterises projects, and set out the important role that narratives play in transitioning between them. Characterised as projecting, they explain how we need to better understand future visioning narratives if we are to understand what underpins project organising. They construct a model of the Un-certain Complexity Complicated Hidden (UnCoCoH) aspects of project realities that links cognition with knowledge, leadership, and behaviours through projectivity, providing a more complete understanding of the ways in which projects play out. In so doing, they reveal the inadequacies of current understandings of projects, and explain how projectivity might bridge them to explicate how we might better shape projects. Following on from Bresnen's earlier contribution, here we can see perspectives from BMR being mobilised in ways that fundamentally challenge the ways in which project delivery is considered, while retaining the domain knowledge that provides the site of its enactment.

In the following chapter, Steve Rowlinson interrogates research that has attempted to address one of the most intractable issues facing the construction industry – occupational safety and health. He explores the various strands and oeuvres of work that have largely failed to make a significant difference to the ways in which safety is managed, or the outcomes achieved. He argues that pre-existing theories have not been applied effectively, but drawn upon selectively leading to solutions which fail to attend to the issues they seek to address. He contrasts the safety and health literature with the approaches adopted within the field, approaches which become intertwined and conflated with other production priorities. Moving forward, the CMR community must surface the tensions that exist between the production imperatives that pervade the industry's occupational health, safety, and welfare, but in ways that tackle the institutional issues that shape current practice. This will, in turn, require a very different research agenda to that which dominates this important field.

In the final chapter in this section, Evangelos Pantazis, Eyüphan Koç, and Lucio Soibelman explore how the potential benefits of new technologies require concurrent and intertwined managerial thinking if their potential is to be realised. They explore how the concept of Lean could help in the uptake of new technologies in construction, or what has become known as industry 4.0. Industry 4.0 has emerged as a label for a whole range of production concerns and techniques founded on smart techniques, simulation/modelling, and digitisation, but the benefits have not been fully realised in construction. Lean, on the other hand, is a managerial philosophy based on value generation theory whereby customer needs are served by a continually improving production process. Although they have their genesis in different drivers and production philosophies, the authors provide a framework for Lean delivery within the 4.0 era, which provides a set of provocations for CMR to embrace. However, they also espouse the need for a concurrent philosophical transformation, a focus often neglected by CMR.

New opportunities for evolving CMR practice

In Chapter 10, we turn to quantitative enquiry and, specifically, the ubiquitous use of questionnaire surveys in CMR. Dominic D. Ahiaga-Dagbui and Igor Martek lament the overreliance of the field on the use of surveys as a 'fast and frugal', but ultimately weak and confirmatory, approach to doing perceptions-based research. Their departure point is to position CMR as residing in a liminal space between what could be considered as engineering and management. Researchers draw from a range of fields and apply these ideas to a construction project context, becoming trapped in a cycle of borrowing ideas and applying them in context. The lens of understanding is, however, predominantly framed by surveys of perceptions rather than through observational insights of actual practices. The rapidly administered survey, while offering benefits in terms of efficiency and speed, has become the dominant tool of choice for much of the community in gathering such insight. The fragilities of questionnaire surveys are often overlooked, and their formulation and application are often flawed relative to fields where human behaviour is the focus per se, rather than part of myriad factors which characterise construction management. The reproduction of poor, inward-looking, survey-based research serves no one, least of all the industry that we seek to understand. Rather, the authors argue for the greater use of alternative datasets, and notably those readily available through new media that might shed light on construction practice in other ways. In contrast with Schweber and Chow, however, they do not see the quest for better theorisation as the key here, but rather the pursuit of research approaches which offer deeper and more profound insights to practice.

Following on from Ahiaga-Dagbui and Martek's call to move away from questionnaires towards new data possibilities to provide improved observations of practice, in Chapter 11, Ninan, Krishnamurthy, and Mahalingam delve deeper into the realms of possibility offered by so-called 'new age' datasets provided by digital platforms. They see this as an untapped resource, and as a readily available and valuable alternative to ethnographic insights which, while providing the depth of understanding, remain a niche subfield within the CMR domain. The authors position the data opportunities afforded by the online environment not just as a window into construction practice (even though it does offer profound insights in this regard), but as a constituent part of construction projects. They also overcome the inherent issues in other data, such as bias, the 'Hawthorne effect', and the at times obtrusive demands on research participants. Sources such as social media, communications, and digital media must, however, be carefully analysed and ethically mobilised in pursuit of research insights. Significant questions remain as to the lines of enquiry best served by such datasets or the best techniques for their interrogation. Failure to consider their rigorous application might see them suffer from the same kinds of limitations that have beset questionnaire surveys within CMR (as set out in Ahiaga-Dagbui and Martek's chapter).

In our final chapter, Dan Sage offers arguably the most challenging contribution in this volume. He questions whether construction needs management at all. He is careful not to offer this as an attempt to close down the field, but rather to ask whether alternative forms of organisation might offer better ways of managing construction and, by extension, more fruitful approaches to consider in research. Sage's point of departure is that the managerialisation of construction is a reflection of the broader managerialisation of work, organisation, and society, and yet the implications of this have been surprisingly neglected by those working in CMR. Drawing upon the work of critical scholars such as Martin Parker, he sets out a counterpoint that illustrates what alternative forms of organisation might offer to the inevitable inequalities that managerialisation affords. He brings this into focus through three case examples of how alternative forms of organisation can yield positive differences for those who reject the dominant orthodoxy. But embracing these alternative paradigms requires that CMR also rejects the normative 'conditions of success', as well as considers what their outcomes might offer for the broader socio-economic landscape. Sage lays out a few of the popular agendas that might be served by such a perspective, from values to discrimination to sustainable construction. In each case, the reader is left to reflect on the very different CMR landscape that might result from a rejection of the managerialist tendencies of our field. Sage's contribution resonates with the theme recurring throughout this volume – that the field needs to look outwards to engage with the broader contexts and fields to which it relates and contributes.

Final reflections

In this opening chapter we have set out a position that the construction management field needs to change its modus operandi if it is to advance the theory and practice of construction management. If our community gets this right, then the bleak future posited by some might give way to a brighter one whereby the field leverages its outstanding access to organisations and projects and great relationships with industry bodies and policy makers to inform and shape both theoretical and practical debates.

It goes without saying that the perspectives mobilised by our authors are ontologically and methodologically diverse, reflecting oeuvres which span the entirety of our field so as not to privilege any perspective over another. In reading these thought-provoking pieces, we encourage the reader to reflect upon the complex relationship between the researcher, the industry context that forms their arena of interest, and the researched subjects or actors that populate it. Within an industry like construction, these relationships are always in flux and subject to constant interplay, woven together in a complex tapestry of ever-changing relations. We argue that viewing the research process itself as a product of this interplay will lead to better outcomes for the research community and the industry than if we merely focus on the binary between rigour and relevance, which has dominated debates in recent years. It is up to the individual construction management researcher to establish a place for themselves in this debate, whatever it reveals.

So there we are. Rather than try to prescribe a way forward, we offer a simple question as an aid to reflection on the future of CMR; what do we want our field to be, and who do we want it to be for? This question is far from trivial, and could, we suggest, be a generational issue. CMR academics can and do play several important roles, not only in acting as a mirror to industry, but in questioning and criticising professionals and policy makers in ways that encourage practitioners to reflect on their own actions and assumptions, and even solving the occasional pithy problem for practitioners along the way. Done effectively, this approach should encourage reflective practitioners to take seriously a broader range of considerations than they might normally consider, and could act as a counterweight to the general tendencies towards self-interest that are so pervasive in commercial organisations. This holds equally true in terms of criticism of policies, policy makers, and the policy process. One of the crucial elements to this kind of research endeavour is to draw attention to the assumptions relating to the consequences that are likely to follow from various actions, the likely effects on relevant agents, and the results, both intended and unintended, of the inevitable trade-offs that exist between a host of competing values, perspectives, and political fashions.

Acknowledgements

The convictions that we have set out in this chapter have been shaped and informed by the outstanding scholars with whom we have worked over the years, some of whom have contributed to this volume. While it would not be appropriate to name them here, we like to think that they will see traces of their influence in what we have written and in the ways in which we conduct our research.

References

Bilge, E.Ç., and Yaman, H. (2022). Research trends analysis using text mining in construction management: 2000–2020. *Engineering, Construction and Architectural Management*, 29(8), 3210–33.

Dainty, A., Leiringer, R., Fernie, S., and Harty, C. (2017). BIM and the small construction firm: a critical perspective. *Building Research and Information*, 45(6), 696–709.

Gosling, J., Towill, D., Naim, M., and Dainty, A. (2014). Principles for the design and operation of engineer-to-order supply chains in the construction sector. *Production Planning and Control*, 26(3), 203–18.

Green, S.D. (2011). *Making sense of construction improvement*. Wiley-Blackwell, Oxford.

Gruneberg, S., and Francis, N. (2019). *The economics of construction*. Agenda Publishing, Newcastle, UK.

Harty, C., and Leiringer, R. (2017). The futures of construction management research. *Construction Management and Economics*, 35(7), 392–403.

Langford, D., and Hughes, W.P. (2009) *Building a discipline: The story of construction management*. ARCOM, Reading, UK.

Leiringer, R. (2020). Sustainable construction through industry self-regulation: the development and role of building environmental assessment methods in achieving green building. *Sustainability*, 12(21), 8853.

Leiringer, R., Gottlieb, S.C., Fang, Y., and Mo, X. (2022). In search of sustainable construction: the role of building environmental assessment methods as policies enforcing green building. *Construction Management and Economics*, 40(2), 104–22.

Leiringer, R., and Dainty, A. (2017). Construction management and economics: new directions. *Construction Management and Economics*, 35.

Pettigrew, A. (1997). The double hurdles for management research. In T. Clarke (Ed.), Advancement in organizational behaviour: essays in honour of Derek S. Pugh. Dartmouth Press, London, pp. 277–96.

Schweber, L. (2015). Putting theory to work: the use of theory in construction research. *Construction Management and Economics*, 33(10), 840–60.

Van de Ven, A.H. (2005). *Engaged scholarship: a guide for organizational and social research*. Oxford University Press, Oxford.

2 Construction economics – it ain't what it used to be

Stephen Gruneberg

The complexity of construction: beyond simple supply and demand

Conventional approaches to construction research tend to rely on a mechanistic supply and demand analysis to explain the operation of construction markets. This assumes that the equilibrium price is the price that occurs when the quantity supplied is equal to the quantity demanded. The market is somehow regulated by the price. If the price is above the equilibrium, then more is supplied than is demanded, resulting in a surplus. Price then declines in response to a surplus, as firms reduce their prices to offload their spare capacity or excess stock. Similarly, if the price is below the equilibrium price, the quantity demanded increases, while the quantity supplied decreases. According to this approach, the quantities supplied and demanded therefore adjust in response to the market price.

Although supply and demand is a very useful concept for understanding markets, this is not an accurate description of how markets in construction work in practice. Below I provide a description of how the construction industry actually functions in economic terms. This description, however, does not fit into a neat theory or model of construction. It only sets out the business context in which the built environment is produced and maintained. It is a mixture of conflict, interested parties, legal disputes, improvisation, disorganisation, and constant change. Piketty (2014) argues that inequalities in the distribution of both wealth and income exist and that these inequalities are not determined purely by economics, because political and historical factors intervene to affect the power of the various parties to a contract. This also applies to the construction industry, where external factors frequently cause unexpected or unanticipated problems that influence the outcome of a project.

The immediate economic role of the construction industry is to create and maintain the built environment and its infrastructure. However, the construction industry also reinforces wider political and economic forces in society. Why, we may ask, if supply and demand is used to meet the needs of society, is the housing market continually underperforming? Why is homelessness a persistent problem? Homelessness remains a problem because the market only caters for those who are able to participate in the market because they have a sufficient income to make their demand effective. Effective demand is the quantity that consumers are willing and able to buy. Those whose income is too low cannot enter the market even if they wish to buy. Their demand is ineffective. Only those who can afford the market price can participate. If there is a shortage in a market, prices will rise in response. A higher price attracts a greater supply, and shortages will disappear, but this will not help those who cannot afford housing in the first place. In the housing market, what actually occurs is a persistent homelessness brought about by house prices being too high for the whole population to afford to buy or rent, too few houses being built to meet demand, and a lack of political will to ensure a basic human need is met when incomes fall short of the market price.

For houses to be cheap enough for young people to be able to afford to purchase a first home of their own, the current owners of housing, often the older generation, would find the value of their property declining, causing negative equity. Negative equity implies the value of their homes would be less than the price they had paid. Bearing in mind there are far more owner occupiers than there are homeless people, finding a solution to the housing crisis is not easily resolved. It is made more complicated by the nature of the housing market. Simplifying their role somewhat, house-builders are speculative developers. They are not, in reality, house-builders at all. To understand how house-builders operate within the housing market, and indeed how most sectors of the building industry operate to a greater or lesser extent, there is a need to refer to the land market and its role in the value of the built environment.

House-builders buy land. They then build housing on it by using small firms (subcontractors) to carry out the building work; finally, they sell the finished houses. They make their profits from buying and selling land, not houses. When land prices rise, house-builders respond in order to take advantage of the rise in land costs. They need to build houses to sell the land at a premium. They purchase land when it is cheap and sell it again when it is relatively expensive. House-builders maintain land banks for this purpose. Land banks are a reserve of potential building sites with or without planning permission. Obtaining planning permission enhances the value of a site, which house-builders use to supply houses, when land – and therefore house – prices are high. Their role as builders of homes is almost a side effect of their business model. Subsidising house purchases only serves to raise prices and increase the indebtedness of buyers in the long run. In the short run, it does nothing to make housing cheaper or more affordable, but simply continues the upward movement in house prices.

Although the discussion thus far appears to be polemical, it demonstrates how the interests of one group are adversely impacted by construction, while the interests of another group may benefit. Conflict and confrontation are central to understanding the construction process and, indeed, any production process. Construction projects embody the power struggle taking place between all the participants: not only between the landowners and the developers, but also between the labourers and their employers, between all competing firms, and between the various materials and technologies involved. Each decision taken affects different firms in different ways, influencing their behaviour and attitudes within the building team, affecting how closely they work together as part of the team or whether they resort to litigation to seek compensation. Under these circumstances, it is hardly surprising that disputes are common in construction projects.

It is also important to take into consideration the nature of construction and how the end product contributes to society. Not all new build increases the stock of buildings and structures in the built environment, because a proportion of new build only replaces buildings that have been demolished. Apart from new build, which includes all kinds of buildings and structures, the remainder of building work is work on existing stock or repair and maintenance, refurbishment, and adaptations to alter the use of buildings. Repair and maintenance of existing stock is an ongoing process until buildings become obsolete. Building obsolescence occurs when the actual or imputed rental income of a building is insufficient to cover the cost of its maintenance, at which point it becomes dilapidated and ever more expensive to repair, and demolition becomes an option. Similar factors lie behind the financial and economic decisions to invest in the built environment, whether new build or work on existing stock, and often depend on external economic changes and developments affecting particular sites. It is these external circumstances that affect the viability of certain building sites, possibly making them more attractive to developers, who may have spotted a new opportunity, or they become less attractive and therefore remain unused. Finally, it is important to keep in mind that demand for construction is a derived demand, derived from the demand for buildings and structures. Indeed, the demand for factories and offices is also a derived demand and depends on the demand for the products or services made by the firms requiring new premises. The source of demand for construction may well, therefore, be distant from the construction site itself, which means that decisions and activities in construction are one or more steps removed from the final customers. These complex characteristics of the sector have important implications for the ways in which it can be understood, and by implication, the ways in which it can be researched from an economic perspective, as this chapter will reveal.

The scope of the construction sector

The process of production defines the construction sector as a series of industrial activities beginning with the extraction of raw materials in the primary sector. Firms in the secondary sector manufacture and assemble buildings and infrastructure. The tertiary sector provides services such as finance, hospitality, and catering. In construction, the tertiary sector covers estate agents, banks and finance houses, and plant hire. The tertiary sector does not make anything as such, but is needed to facilitate the production process. In reality, the construction sector embodies all three sectors of the economy in an integrated whole. The scope of the construction sector is therefore more complicated than it first appears. It is far greater in scope than what occurs on a building site. It includes contractors, specialist building firms, and their supply chains, including a multitude of construction products produced in quarries, factories, and design studios.

Construction activity exists within every economy in the world. How it manifests itself in each country depends on the history, geography, politics, and culture of each land. Each country has a construction industry that reflects the characteristics of that country, its laws, and traditions. Taking the UK as an example, the construction industry extends to trade associations and professional bodies, such as the Royal Institution of Chartered Surveyors (RICS), the Royal Institute of British Architects (RIBA), and the Chartered Institute of Builders (CIOB), who play an important role in the workings of the industry. The one thing these professional bodies have in common is that they are employers' organisations. Trade unions represent the labour side. Following years of mergers and amalgamations, two unions emerged as the dominant players in the construction industry: Unite the Union, which emerged out of the Transport and General Workers Union, and the GMB Union, which grew out of the General, Municipal, Boilermakers and Allied Trades Union. Equivalent bodies are to be found in most countries, but they differ in size, rules, and practice. These differences emerge as a result of the unique history and power structures in each country.

Professional bodies and trade unions are part of the institutional context of the construction industry. The term 'institution' covers all the structures and relationships that combine to form the culture of an industry and can be referred to as the social structures of accumulation, as they all contribute to the production process in one way or another. This refers not only to organisations in the construction industry, but also to all sources of information and data used by firms to inform their business decisions, and includes particular practices that firms expect to follow in their dealings with other firms in the industry, such as organising auctions to find the lowest-cost subcontractor. Gruneberg and Ive (2000) discuss the social structures of accumulation to frame research into construction, and argue for an approach that takes into account the social, political, and economic factors that lead to an understanding of how firms behave in different markets. For example, every specialist trade has its own characteristics and, as a result, there are institutional and cultural differences between the many trades. These characteristics determine the practices of firms in different specialisms and markets. Even within construction, the behaviour of firms varies from one trade to another and from one time period to another. These changes over time describe the changes taking place in the way firms do business, for example, as a consequence of the arrival of new technology.

When it comes to defining what is included in the construction sector and what is not, the economics approach is to look at the statistics. Again taking the UK as an example, all firms in the UK economy are allocated a Standard Industrial Classification (SIC) code to enable data to be systematically collected. These data are based on the value added by every industrial activity in the economy to calculate the national income and the contribution of each type of firm. Table 2.1 shows Section F (Construction) of the SIC, which is subdivided into SIC two-digit codes: 41) Construction of buildings, 42) Civil engineering, and 43) Specialised construction activities. Together they measure construction activities on site, including site preparation, the building of complete structures, plumbing, roof covering, plastering, and wall and floor covering. But this is not a full list of construction activities, and different kinds of activity that are connected to construction can be found in other divisions of the SIC. For example, Section C covers manufacturing and includes the manufacture of builders' carpentry, cement, insulated wire, and cable; Section J covers financial intermediation, including building societies and mortgage finance companies; and Section K includes real estate, renting, and other business activities that are relevant to the building industry. Several other activities do relate directly to the built environment, but are included elsewhere in the statistics, such as quarrying for slate, stone, and gravel and manufacturing builders' carpentry, glass, bricks, cement, central heating radiators, cooling and ventilation equipment, and lighting equipment.

Table 2.1 Standard Industrial Classification for construction based on SIC (2007)

Construction	
41	**Construction of buildings**
41.1	Development of building projects
41.10	Development of building projects
41.2	Construction of residential and non-residential buildings
41.20	Construction of residential and non-residential buildings
41.20/1	Construction of commercial buildings
41.20/2	Construction of domestic buildings
42	**Civil engineering**
42.1	Construction of roads and railways
42.11	Construction of roads and motorways
42.12	Construction of railways and underground railways
42.13	Construction of bridges and tunnels
42.2	Construction of utility projects
42.21	Construction of utility projects for fluids
42.22	Construction of utility projects for electricity and telecommunications
42.9	Construction of other civil engineering projects
42.91	Construction of water projects
42.99	Construction of other civil engineering projects nec (not elsewhere classified)
43	**Specialised construction activities**
43.1	Demolition and site preparation
43.11	Demolition
43.12	Site preparation
43.13	Test drilling and boring
43.2	Electrical, plumbing, and other construction installation activities
43.21	Electrical installation
43.22	Plumbing, heat, and air conditioning installation

Construction	
43.29	Other construction installation
43.3	Building completion and finishing
43.31	Plastering
43.32	Joinery installation
43.33	Floor and wall covering
43.34	Painting and glazing
43.34/1	Painting
43.34/2	Glazing
43.39	Other building completion and finishing
43.9	Other specialised construction activities
43.91	Roofing activities
43.99	Other specialised construction activities n.e.c
43.99/1	Scaffold erection
43.99/9	Specialised construction activities (other than scaffold erection) n.e.c

The statistical definition of the construction industry is given in Section F (Table 2.1). This may be thought of as the core of the construction industry. However, as shown in the paragraph above, this narrow definition of construction omits the many firms and professional practices needed to complete the supply chain of construction, such as quarries, building product manufacturers (including brick manufacturers), architects, surveyors, and estate agencies. A broad definition of construction would include all firms that contribute a significant proportion of their output to the building industry in one way or another. Although they contribute to the finished output of the building industry, several specialised construction-related activities are excluded from the statistics of the construction industry. As a result, their data are not included in total construction output, and the size of the construction industry is often understated, even in official reports. The reason for this apparent statistical anomaly is that the same data are used for several purposes. For example, one use of the data is to categorise industries in terms of sectors of the economy, namely, the primary sector (agriculture and extractive industries), the secondary sector (manufacturing industries), and the tertiary sector (services). Construction extends across all three sectors, as do many other industries. Another purpose of the data is to establish the value added by each specialist industry, rather than the total size of that particular sector of the economy.

Another reason for the undervaluation of construction-related activity in the economy is because the data in industrial statistics are based on value added. Ive and Gruneberg (2000) point out that the construction industry is larger than the UK Office for National Statistics (ONS) figures state, because the data are based on value added on site to avoid double counting when calculating the size of the UK economy as a whole. Omitting certain industries from total construction output may affect policy and economic performance but be perfectly consistent with the aim of establishing the contribution of the construction industry to gross domestic product.

A number of implications flow from this paradox of excluding firms in the wider construction supply chain from the building industry data, even though they contribute to the sector. For example, investment in construction may have a multiplier effect, based on the ratio of the aggregate increased income over the original investment. The circular flow of income describes how income flows round the economy in ever-decreasing amounts. As the income passes from one individual to the next, some income is syphoned off in the form of taxation and savings. Eventually, the total increase in income as a result of an investment is several times greater than the original investment that initiated the process. If the multiplier effect of a given investment was, say, 2.54, according to the theory, an investment in construction of £100 million would eventually be worth 2.54 x £100 million or £254 million to the economy as a whole. Ernst and Sarabia (2015) describe the multiplier effects in a number of high-income countries in different years: 3.92 in 1992, 3.45 in 2006, and 7.74 in 2009, compared to 3.63, 3.06, and 3.45 in middle-income countries in the same years, respectively.

Now, assuming the multiplier is based on the total cost of a construction investment, the value of the investment should cover the broad definition of construction. When carrying out research, the question is whether or not investment in construction includes all construction firms, or only those included in the narrow definition given in the Construction Statistics Annual, which includes only those specialist contractors working on site. As described above, the official ONS data on the construction industry use the narrow definition of construction to provide the annual value of construction output produced on site by type of work, the range of construction firms by trade, and the total number of people employed by each trade. This narrow definition of the construction industry can be used to make international comparisons, as similar data exist for many other countries in their own national datasets, including the USA, Australia, and the countries of the European Union. National data can also be used to show the relative size of different aspects of the industry. For example, the graphs in Figure 2.1 show the breakdown of output by building type, the share of building work by trade of firms, and the variety of construction employment by skills, trades, or crafts in 2019. Comparing differences in the time series of these variables can be used to show the impact of government policies and the introduction of new technology and techniques. A time series can be easily devised by using the equivalent annual figures of a variable for consecutive years.

Figure 2.1 measures construction output in the public and private sectors according to the type of work, including new housing, infrastructure, hospitals, and commercial buildings, based on those firms working on site. This data set indicates the relative size of the markets for various building types. For example, in 2019, according to the ONS data, new housing comprised 39.50 per cent of all new build work, education building work constituted 8.97 per cent, and offices were 8.70 per cent. These measures can be used to compare construction industries internationally. Much information can be gleaned from the data, including the relative size of different construction markets and differences in growth rates of equivalent construction markets in different countries. This can give an indication of spare capacity in the construction industry in different markets by comparing the size of the market in any one year compared to recent peaks in output.

Figure 2.2 shows the output of key construction trades. These data can be used to establish the annual growth rates of different specialisms. By gleaning the value of each trade in each year at constant prices, a time-series analysis of each trade could be used to show changes over time and the rate of growth of each specialism to indicate changes in the composition of the construction industry to derive the structure of that industry. Again, international comparisons can be drawn to compare growth rates of different trades in different countries. This may account for differences in productivity in the construction industries of different countries, just as the different types of structures may also account for differences in productivity across countries. Of course, it is always the case that statistics can be criticised for the inaccuracy of the data, but without claiming the statistics to be accurate, they are nonetheless the best estimates available. It is still possible to state categorically that the figures are derived from official published data.

Figure 2.3 illustrates the range of skills employed by the construction industry. The largest trades in construction in 2019 were the installation of electrical wiring and fitting, at £19 569 million, followed by plumbing, heat, and air conditioning installation, at £13 469 million. These data are published annually, enabling changes in employment in various trades to be shown over time, bearing in mind that the value of each trade is at current prices. Taken

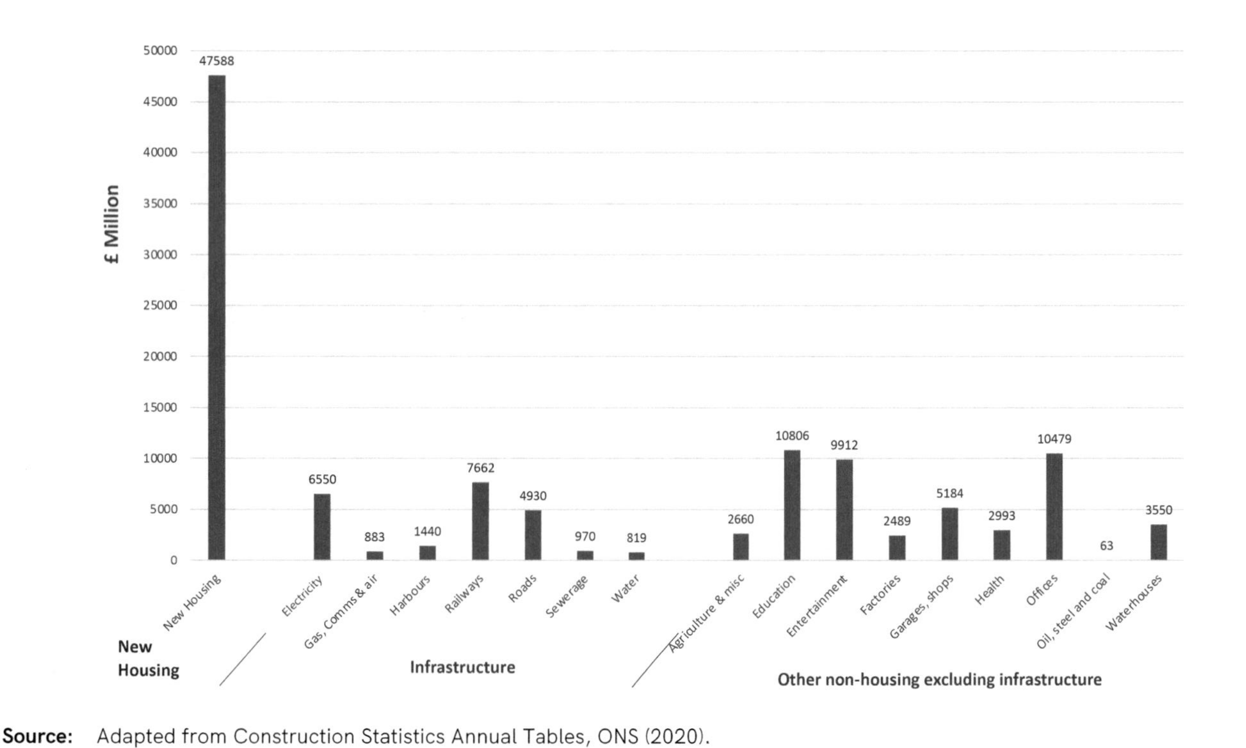

Source: Adapted from Construction Statistics Annual Tables, ONS (2020).

Figure 2.1 Value of construction new work output, by type of work in the UK, 2019

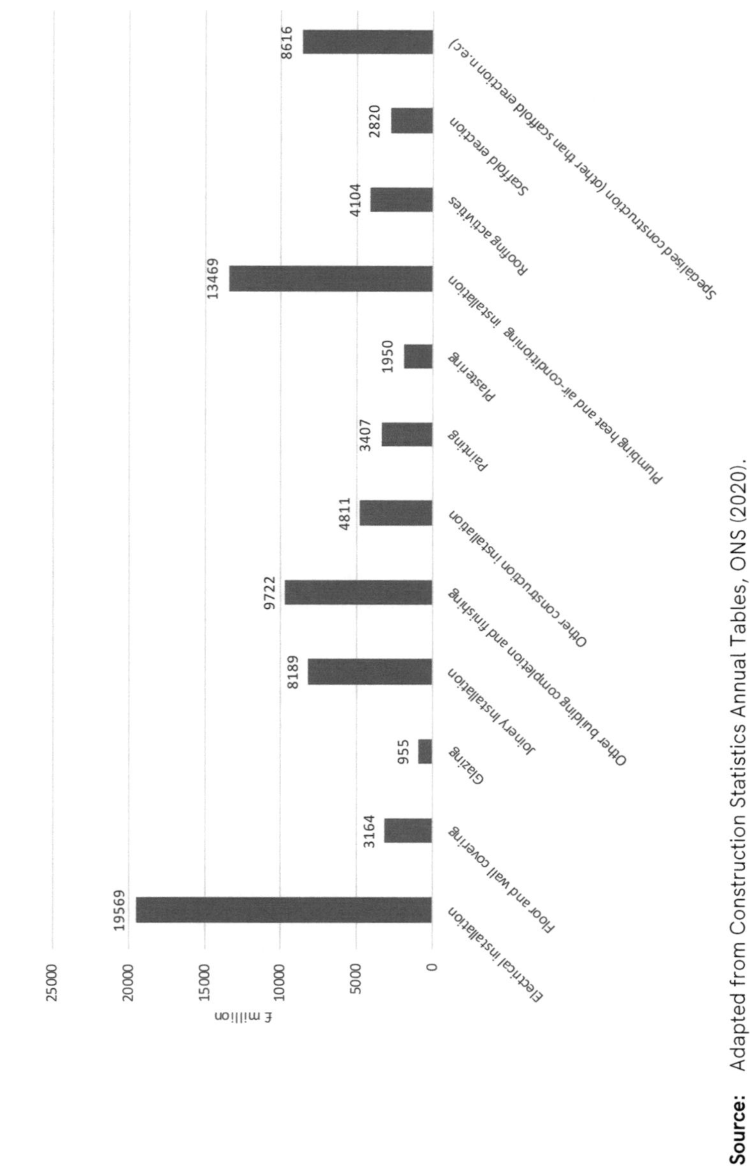

Source: Adapted from Construction Statistics Annual Tables, ONS (2020).

Figure 2.2 Construction firms: value of work done, by trade of firm and type of work in the UK, 2019

together, Figures 2.1, 2.2, and 2.3 show how fragmented the construction industry is in terms of the different buildings and markets it serves. This is very useful for establishing the reasons for differences in construction productivity in different countries as different trades use different skills and amounts of capital.

How construction markets work: auctions

The conventional approach in most economics textbooks assumes that in each market there is a demand for a given quantity at any given price and a willing number of suppliers willing to meet that demand. At the equilibrium price, the quantity demanded is equal to the quantity supplied. At prices above the equilibrium price, surpluses occur because more is offered for sale than buyers are willing to purchase. Similarly, at prices below the equilibrium, buyers are willing to buy more than sellers are prepared to offer for sale. As a result, shortages appear. Unfortunately, in the construction industry there is little empirical evidence to support this textbook approach. In reality, although there is competition between sellers to win orders, this competition takes the form of an auction in which there is usually only one winner, and the winner takes all. In construction auction markets, contractors are invited to bid or tender for work. The buyer is therefore in a strong position to set one seller off against another to obtain the best (usually lowest) price. Tendering for work in this way occurs at every level in the construction process. Main contractors bid against each other to be appointed by the client. Specialist subcontractors compete with each other to be selected by the main contractor in similar monopsonistic markets. This method of selecting suppliers tends to attract the lowest bids as firms compete on price rather than on quality, ability, and skills, leading to consistently low profit margins.

In construction, a firm's portfolio of projects is undertaken as a series of separate orders. Starting at different times, projects are carried out sequentially, though they may be carried out at the same time as other programmes, with some overlapping occurring. Each project is bespoke and has a unique combination of characteristics in terms of size, design, location, contracts, and participants. This is a very different notion from that of a market for a homogeneous product or service for which there is an established demand and supply, and it has profound implications for how we can understand the operation of construction firms, for example, their willingness to invest in new technology and the training of staff.

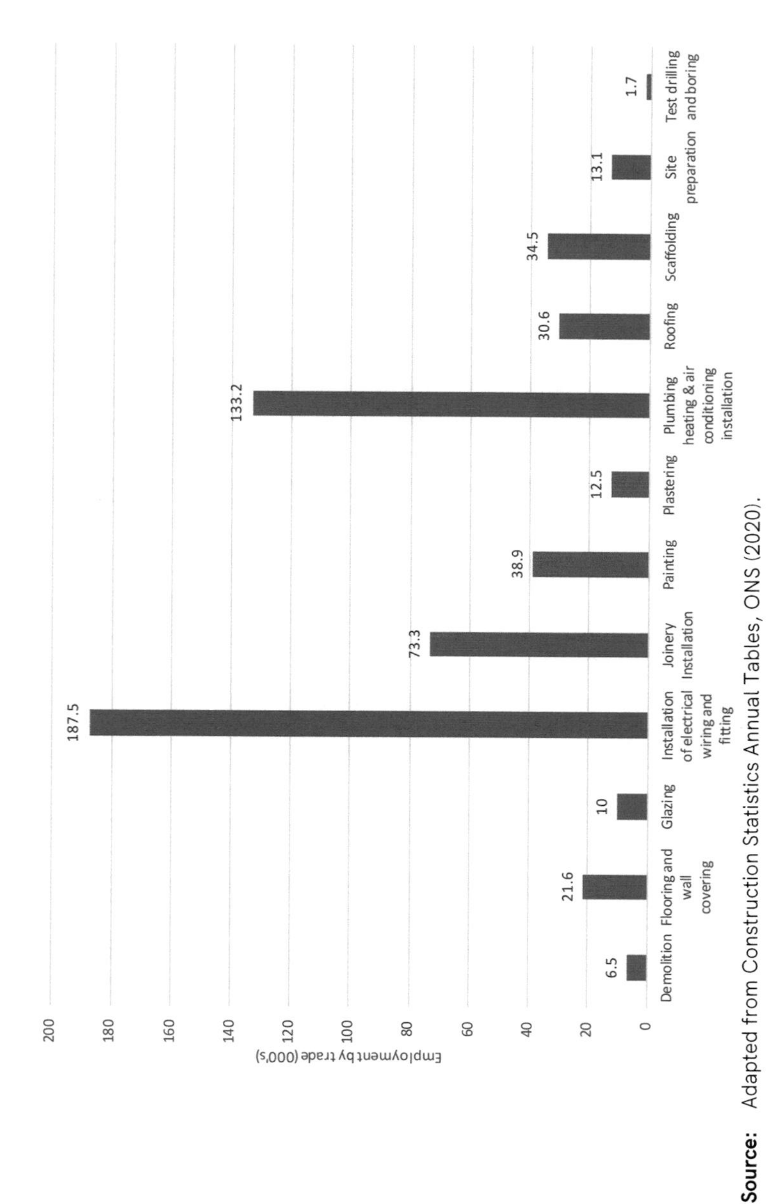

Source: Adapted from Construction Statistics Annual Tables, ONS (2020).

Figure 2.3 Total employment in construction, by trade in the UK, 2019

In the tendering process, each contractor has to price the work and add a mark-up for profit. Because contractors take on projects sequentially, at the beginning of the year of trading, the contractor has to select a mark-up on costs without knowing how much is needed to cover overhead costs, such as head office costs and annual tendering costs. If the mark-up is too low, the contractor could have charged more and still have been the lowest bidder for the work. This is known as leaving money on the table. In other words, they under-priced their own output. If the mark-up is too high, they will have priced themselves out of the market unnecessarily. They have to tender for work before they know what a reasonable mark-up might be, giving them a level of pricing needed to win sufficient contracts to pay for the annual overhead costs of tendering to keep the firm fully occupied, or to avoid taking on so much work that delays occur and penalties for late delivery are incurred. Other practical difficulties arise when contractors find that the costs of materials and labour rise over the period of the contract, if demand across the industry rises unexpectedly during the period of the contract. These uncertainties can lead to losses on projects, which firms can ill afford due to their low profit margins, which leaves little room for error. The risk of losses on a project leads contractors to find ways of cutting costs whenever the opportunity arises. This in turn may lead to mistrust, disputes, and litigation.

Prior to contracts being agreed and signed, the supplying main contractor has little influence over the size and design of a project or even the materials used and building methods. Contractors have complained for many years that they are brought in to advise after the design has been decided and it is too late to influence the project to take advantage of the contractor's skills and experience. It only remains for the builder to decide whether or not to accept the proposed design, by which time it is too late to influence the design to improve its buildability, taking advantage of the contractor's technical knowledge and experience. However, once a contract has been agreed and signed, negotiating power shifts from the developer to the main contractor. Replacing a contractor after work has begun on site is expensive for the developer and risky for the new contractor. Indeed, the management of risk plays a central role in the management of projects. One of the purposes of subcontracting is to allocate work packages to firms that can take on the size of risk involved. As firms are constantly seeking to offload risk onto others, it is important not to burden a supplier with financial risk it cannot bear or afford, because the failure of a subcontractor to meet its obligations only increases costs for everyone involved in the project. Herein lies plenty of scope for future research, as the implications of risk-allocation strategies not only have direct consequences for the project in question but also create ripple effects that can have far-reaching consequences on construction markets.

Of further importance in understanding firm behaviour in construction is how individual companies seek to maximise their profits. The standard economic-theory textbook account of a profit-maximising firm assumes the firm is in a position to decide the quantity it produces and the price it charges. This is not the case in construction because the contractor is not in control of the size of the project or its design characteristics. Unlike other industries, contractors are not usually in a position to decide how much output to produce, such as the floor area or the number of storeys. Nor are contractors in a position to decide how large a project should be to generate a greater return for themselves. Instead they are contractually obliged to meet the developer's requirements (and planning constraints), rather than meeting their own interests, which comes at a cost in the form of a lower return than if they had been able to behave more selfishly like other profit-seeking enterprises. As a result of low profit margins, contractors have to take on exceptionally large quantities of work in order to generate large enough profits to survive.

Finally, project procurement is largely concerned with matching work packages with particular specialist contractors in terms of specialist knowledge, skills, and financial ability to take on the risk. Allocating work to different subcontractors is one of the main roles of main contractors. Each participant in the construction project team is part of a complex supply chain, extending from the main contractor, through to each specialist contractor and the myriad of construction material and component suppliers. All this might come across as relatively self-evident, but the implications are far from trivial. For example, it serves to explain why 'new' approaches such as 'modern methods of construction' and 'design for manufacture' are being heavily promoted in some quarters, but as of yet have seen limited implementation in most countries. An economics lens would have much to offer to research in this area. As will be further discussed below.

The role of innovation in the building production process

Construction management research is about studying all aspects of the building process, with a view to finding new applications of technology and management methods to develop a broader understanding of the built environment and to facilitate the building process. Over the past 20 years, significant research attention has been given to innovations, both technological and those related to the processes involved. At least part of this interest comes from the constant pressure on contractors to reduce the share of the cost of construction relative to the total cost of a building. Hence, while innovations may be used to add value while leaving costs unaffected, the pressure to reduce costs is one of the core incentives to innovate. Looking at innovation from an economics perspective provides a different understanding to that provided in the, mostly technology-centred, construction management literature.

The value of land depends on the capitalised rental value of a site, which is realised when buildings are completed and rented out. The higher the capitalised rental value, the more a developer can pay for a site, allowing for the cost of construction. The value of a building (or a site) has to do with the value to the owner of the building through expected rental income and not construction costs. If a contractor adopts a cost-cutting innovation, the benefit of the building cost reduction can be shared between the developer and the contractor in the short term. However, in the longer term, the gain is passed on to the landowners, as developers compete with other developers to purchase more sites, knowing the cost of construction has been reduced. The benefit of the lower cost of construction is thus passed on to the landowners selling sites, as they receive higher bids from developers. None of this is immediately apparent, but it is one of the reasons lower costs of construction do not reduce the cost of buildings in the long run, a point which is often conveniently forgotten in a large number of policy documents.

The construction sector is often thought of as backward, but this is a misrepresentation of the construction industry and ignores the degree and pace of innovation and change that is taking place. Innovation is taking place continually on construction sites throughout the industry, in all specialisms. However, because the building industry is highly fragmented and the vast majority of firms are small or medium-sized enterprises, the vast majority of innovations are on a relatively small scale. Moreover, there can be up to 30 crafts, skills, and trades working together on a project, and an innovation in any one specialism will only affect a small percentage of the total cost of construction. These small changes are nevertheless taking place within the majority of firms on site. Innovation is, indeed, such a common phenomenon that it occurs almost unnoticed as part of the building process. It is almost taken for granted. Researchers into construction innovation in general, and the adoption of new technology in particular, would do well to treat this as a starting point.

Understanding the nature of conflict in the building process

To further understand innovation and development in general in the construction industry, it is necessary to look more closely at its inner workings. At the heart of capitalist production is the need to generate a surplus, because over the long run surpluses enable firms to survive. Generating surplus value is the motivation behind investment decisions. The transformation of raw materials into finished goods – in our case, completed buildings and infrastructure – enables firms to charge more than the cost of labour and materials. Value added is the difference between the selling price of the finished product or service and the cost of material inputs together with the cost of labour. The share or distribution of value added between the firms in the construction supply chain depends on the power and the ability of firms to negotiate with other construction firms. If the price of the finished output were only equal to the cost of materials, there would be nothing left to pay wages or make a profit, which is the residual after all other costs have been paid. Although firms may borrow to invest, investment is ultimately derived from profits, and without value added and surpluses, there could be no investment in the future.

The above is important because profits and surpluses are the incentives that motivate firms to produce. Firms can only survive if the return on investment is sufficiently high to compensate for the risks incurred in taking on a project or part of a project. Incentives are used to encourage certain behaviour and discourage other behaviour by increasing or decreasing the reward for the appropriate action. Indeed, understanding the motivation behind business decisions may lead to new ways of predicting how firms might react to different government policies, and the lack of incentives may explain the failure of others. For example, just because it may be in the best interests of the various specialist firms to collaborate with their fellow subcontractors working on the same project to take advantage of the mutual gains available by adopting new technology or new techniques, it is not necessarily the case that firms choose to co-operate with the other members of the construction supply chain, especially when delays occur or additional costs are incurred. This branch of economics has come to be known as 'game theory' because it resembles the behaviour of players of board games, cards, or even sporting situations. One such game taken from game theory is 'the prisoners' dilemma', which demonstrates that it is not necessarily the case that firms will collaborate. The prisoners' dilemma is a classic case used to demonstrate that, even when it is in their best interests to work together, the incentives are such that firms do not always co-operate. To illustrate this point, let us imagine that two prisoners, Prisoner A and Prisoner

B, have been arrested for a crime. They are prevented from communicating with each other. Each prisoner is presented with an identical choice. If they both confess, both receive three years of punishment in prison. If neither confesses, they both receive imprisonment for two years. However, if one confesses and the other does not, then the prisoner who confesses is set free, but the one who does not confess receives five years in jail. Both prisoners have to decide whether to confess or not.

While it is in both their interests not to confess, if they do not trust each other, the temptation for each prisoner is to confess, as this would risk, at most, a sentence of three years. However, if Prisoner A, for example, does not confess, Prisoner A then runs the risk of a five-year prison sentence if Prisoner B decides to confess to the crime. The reward for not confessing depends on what the other prisoner decides to do. Because of the lack of trust between them, both prisoners are most likely to choose to confess. This would leave both of them worse off, with three years in prison each, rather than the two years, if only they had both been able to trust the other and not confess.

The prisoners' dilemma reflects the inherent conflict in the construction production process together with the lack of trust between building firms. We have to recognise that there is conflict in any process that delivers a zero-sum game, which is precisely what occurs on construction projects. In game theory, zero-sum games arise whenever the gains of one party are at the expense of one or more of the other players. This occurs in construction projects where there are a limited number of players and all are competing, hoping to increase their own share of the project costs, leaving less for the other firms working on the project. For example, the larger the share of costs going to the main contractor, the less goes to the specialist subcontractors; the more going to the developer, the less remains for the main contractor; and the more going to one subcontractor, the less remains for the other firms.

Often the above-described conflict within the supply chain is ignored by construction management researchers seeking to transform the way the industry operates through the adoption of new technology. Whenever a new technology is brought to site, the new method will benefit some members of the construction team more than others, and sometimes the changes introduced can be disruptive for some, replacing labour, plant, or materials. For example, Building Information Modelling (BIM) allows firms to work closely together using information technology (IT), but the introduction of BIM means that some firms are in a position to increase their share of the contract sum. Moreover, one of the reasons for the introduction of BIM is the possibility of cost reductions that would benefit the developer, the client, and the supply chain in the short run. However, as we have seen, in the long run, the benefits will flow to landowners at the expense of developers and contractors. As prices do not decline, there is no increase in sales overall and therefore no increase in turnover as a result. Indeed, any increase in demand for construction is the result of changes taking place elsewhere in the economy. The reason firms undertake innovative techniques is to compete with other contractors and win particular projects. This contrasts with the practice and experience in other industries where innovations, such as battery technology or advances in photovoltaics, reduce prices and increase demand. From a contractor's point of view, the role of innovation in construction is, therefore, generally limited to individual projects rather than changing the industry. Where innovation may be seen to have a wider impact in construction is with the construction product manufacturers, who compete with other product manufacturers to gain market share. It is through innovations by competing construction product manufacturers that the construction sector benefits from improvements made by construction component producers. This requires ongoing research into changes in construction processes brought about by technological developments in firms, not necessarily construction firms, who invent and innovate by introducing new methods and products to contractors engaged in the building process.

Future construction economics research - a call to arms!

In many ways, the construction industry is a microcosm of society. The industry is fragmented into a large number of specialisms, markets, trades, and skills. Firms are appointed prior to work commencing; therefore, there is a great deal of risk and uncertainty facing everyone involved in the process. It is not surprising that many disagreements arise during the construction phase. Studying the economics of construction is like untangling a large number of intertwined knots. Researching aspects of the construction industry is invariably an intellectual challenge.

In classical economics, it is assumed that markets tend to return to equilibrium when all economic forces are balanced and there is little need to intervene. In construction, this is not the case. Markets tend to favour the stronger and larger companies at the expense of the weaker firms and unorganised workers. Wealth and income is concentrated in the hands of the most powerful participants. Relying on market forces to resolve disputes and difficulties only serves to reinforce an unequal balance of power. Hence, there is a need for government intervention and public sector policies and strategies to deliver a more equitable distribution and resolve disputes. In the construction sector, the challenge for many firms is survival itself, in the face of low profit margins, very high sales volumes, and highly competitive markets. Construction management researchers looking to examine the improvement of individual parts of the construction process do well to remember that it is a high-risk industry.

The selection of firms is invariably based on auction markets, where the lowest bid is used as the criterion for choosing building contractors and specialist construction firms. This tends to create a downward pressure on prices and profit margins, forcing firms, in turn, to put further pressure on their suppliers to reduce their prices. It is important to emphasise the significance of the extremely low profit margins in the construction industry. Because of the low profit margins, there is little slack to employ people for longer than they are needed on site. Training is minimal, and terms and working conditions tend to be poor relative to other industries. Yet, the construction industry's economic environment is seldom discussed. Instead of understanding the issues confronting contractors and subcontractors, those left to deal with the problems of this hazardous, disorganised bear-pit of an industry – namely, the contractors – are often deemed uncaring, untrustworthy, and litigious. They are blamed for the failures of the construction industry: low productivity, cost over-runs, late completion, and poor quality.

It does not have to be this way. A more positive understanding of the economic plight of firms working in construction, underpinned by rigorous research, could deliver an awareness of the importance of training, a concern for the reputation of the industry and the firms that contribute their productive inputs, and bring about a cultural change altering the social structures of accumulation mentioned at the start of this chapter. Unfortunately, as things stand, there are few incentives for contractors. For example, if a contractor develops a new process, material, or technique, there is every chance it will be copied by competitors, undercutting any price incentive the original contractor might have hoped for. There is a large gap in training in the construction workforce, as trained staff earn more than unqualified workers and contractors see their trained workers leaving for jobs in competing firms. The Construction Industry Training Board (CITB) in the UK and the many similar initiatives around the world are valiant attempts to overcome these difficulties, but as has been shown time and time again, more resources are needed to overcome the gap between the need for training and the cost of training. Once again, the economics of the market undermines the effort in this direction. Only when there is a shortage of trained skilled workers do wages increase. It is a paradox of the labour market that shortages of labour arise when the wages on offer are too low to attract skilled workers. But if incomes were high, then the abundance of labour would form a downward pressure on wages. A construction economics research perspective on these labour market issues would lead to a better understanding of the contradictions of the construction labour market and why some labour policies are ineffective, while other research may suggest the kind of actions/initiatives that are needed.

The key argument I have progressed throughout this chapter is that, if the issues confronting the construction industry are to be understood, it is necessary to consider why the industry functions the way it does and what might be done to improve it. This, I contend, affects construction management research at all levels, and numerous examples can be provided ranging across the entire construction process (i.e. planning, design, construction, and use). As I have argued above, one reason the industry is so confrontational is because it is fragmented into many specialisms, trades, and crafts. Equally important are the low profit margins in construction that force firms to cut corners and that render them unable to make concessions in any dispute that may arise. This leads to an atmosphere of distrust and opportunism, often referred to as a 'dog-eat-dog' culture in which firms take advantage of each other whenever they can. The threat of litigation is real, as there is little opportunity for firms to establish close working relationships with firms they only work with occasionally. Therefore, little business turnover is lost through resorting to legal disputes if it is unlikely that any one firm will work with the same firms again. Because of the uncertainty and complexity of the economics of the construction sector, there are many difficult and uncomfortable issues that will always need to be studied, such as the part played by conflict in the building process, the use of economic power in the construction industry, and the use of incentives to encourage firms to act in productive ways in the construction sector. Other fruitful areas of study lie in the details of the construction sector, such as research into specialist contractors' markets. It is important to consider that each specialism in construction exists within its own social structure of accumulation, with its own culture, rules, and institutions, often quite separate from the rest of the building industry.

Broadening the scope of construction management research, the following, based on Gruneberg (2019), suggest a number of strands of thought that might be used to overcome the difficulties of organising an effective, efficient, productive, quality built environment and construction industry by researching and reforming the production of the built environment. This is by no means a comprehensive list of areas for future research, but together they form an agenda for research into the economics of the construction industry:

1. The implications of continuing the development of IT and encouraging its many applications in construction. Adopting a construction economics

approach in this area of research would emphasise the impact of IT on productivity in the construction sector and show how the use of IT would benefit the various interested parties, including land owners, construction component suppliers, building users, and the managers of the building process. However, not all participants in the building process would be affected to the same extent. Research in this area would also need to take into account the distribution of the various benefits to those gaining as a result, and the costs of innovation to the losers, including considering how various participants in the process could be adversely affected by the adoption of radical new approaches offered by the new technology.

2. The longer-term consequences of professionalising the workforce and skills in construction. Construction economics research would enhance a professional ethos in the building sector that would highlight both the importance of the quality of the actual construction output, and the impact such innovation would have on the terms and conditions of employment of labour, its training, and its attitude towards clients and building users.
3. Research into modelling a national plan for the built environment would need to take into account the advantages and difficulties of achieving continuity of work and steady demand for construction. However, a policy of steady construction demand may or may not meet the needs of the wider economy. Using construction economics research to facilitate the planning of the built environment could assist in co-ordinating an overall approach to the building process. The implications of this area of research could demonstrate how improvements in construction methods, assuming continuity of work, would add to the economic growth of the economy as a whole.
4. Another area of research involves studying the reform of the planning system to find where improvements might be made to reduce delays caused by the planning system itself and, hence, speed up the process and save a great deal of time and cost, even before contractors are engaged.
5. Research into proposals in the built environment that would generate zero carbon would ensure the construction industry is seen as contributing towards creating a sustainable economy. It is becoming increasingly clear that many intangible costs and benefits affect the economic viability of projects. This area of cost–benefit analysis has been under-researched in the past, an omission that has led to the built environment being seen as a major polluter.

Finally, one of the greatest single research challenges facing construction researchers is the role the construction industry could have in dealing with climate change and how the construction sector might manage the issue of global warming and the sustainability of both the built and natural environments. This calls for a multidisciplinary approach where construction economics has an important part to play.

References

Ernst, C., and Sarabia, M., (2015), *The Role of Construction as an Employment Provider: A World-Wide Input–Output Analysis*, Employment Working Paper No. 186, Employment Policy Department, Geneva, International Labour Organisation.
Gruneberg. S., (2019), *A Strategic Approach to the UK Construction Industry*, Abingdon, UK, Routledge.
Gruneberg, S., and Ive, G., (2000), *The Economics of the Modern Construction Firm*, Basingstoke, UK, Macmillan.
Ive, G., and Gruneberg, S., (2000), *The Economics of the Modern Construction Sector*, Basingstoke, UK, Macmillan.
Office for National Statistics, (2020), *Construction Statistics Annual*, Newport, UK. https://www.ons.gov.uk/businessindustryandtrade/constructionindustry/datasets/constructionstatisticsannualtables
Piketty, T., (2014), *Capital in the Twenty-First Century*, Cambridge, MA, Belknap Press.

3 What are we trying to achieve? Rigour, relevance and modes of knowledge production in construction management research

Mike Bresnen

Introduction

How rigorous and relevant is construction management research (CMR), and what does it need to do to ensure it maintains the highest standards of rigour and relevance? These are questions that have begun to have a significant impact on discourse within the field through the pages of its leading journals (Dainty and Leiringer, 2019; Harty and Leiringer, 2017; Leiringer and Dainty, 2017a, 2017b; Chan, 2020). Central to the debates these questions provoke is the dual concern that CMR is too selectively engaging with concepts and theories 'borrowed' from business management research (BMR) (Schweber, 2015; Fellows and Liu, 2020) while, at the same time, losing sight of the importance of practical application (Koskela, 2017). In this chapter, I draw upon a similar debate that has raged within the BMR community, emphasising the additional challenges facing CMR in engaging with a disparate, fluid, and contested management knowledge base (Bresnen, 2017). Exploring the different epistemic practices and modes of knowledge production that can shape business and management knowledge, I argue that ensuring greater cross-fertilisation of ideas between CMR and BMR communities of practice is vital in helping promote rigorous and relevant research, but that it requires profound changes in practice at three levels: the institutional, the relational, and the individual. Examining each of these in turn leads to practical suggestions about how to effect such transformational change.

Achieving rigour and relevance in research has been a recurring theme in debates within the BMR academy on both sides of the Atlantic over the last 20 years. Triggered by concerns about the need to combine scientific validity with practical application, several important touchstone works have emphasised the importance of ensuring both rigour *and* relevance in BMR (Pettigrew, 1995; Van de Ven and Jonson, 2006; Van de Ven, 2007). Institutions within the academy, including major international conferences and journals, have given impetus and space to explore how scholarship can become more 'engaged' (Van de Ven, 2007) and how high levels of both rigour and relevance can be simultaneously achieved (Huff, 2000; Hodgkinson, 2001; Bartunek, 2007; Rynes, 2007; Shapiro, 2016). This debate has highlighted the importance not only of transferring or translating knowledge from academic research into practice, but also of transforming the nature of management knowledge itself (cf. Carlile, 2004). Central to this has been an emphasis on the co-production of management knowledge at its source by academics and practitioners, through various forms of collaborative research (e.g. Shapiro et al., 2007).

This emphasis on knowledge co-production has dovetailed with another important strand of thinking originating in the science policy field that emphasises the increasing significance of a suggested new mode of knowledge production whose agenda, aims, and methods are shaped directly by practitioners and their needs (Gibbons et al., 1994; Nowotny et al., 2001, 2003). Gibbons et al. (1994) called this a 'Mode 2' form of knowledge production, which they contrasted with a traditional, more academic-driven approach they labelled 'Mode 1'. While those labels have been refined over the past 20 years, the principles they represent of more practitioner-led problem definition and solution have stuck; and the practices they reflect and promote continue to shape the research funding landscape for BMR, particularly through the heightened importance attached to research 'impact' and the corresponding privileging of some forms of research over others.

How do these wider debates within the business and management field feed into and affect CMR? In this chapter, I explore that question by situating CMR within those wider debates about the challenges of generating research that is both scientifically rigorous and practically relevant. Taking a critical review approach, I unpack the notion of different modes of knowledge production as they apply to CMR, and surface some of the institutional enablers and impediments to flows of management knowledge between business/management disciplines and CMR and between those academic fields and the domain of practice. In doing so, I will draw out implications for CMR moving forward, with particular emphasis on inter-disciplinary opportunities for learning and overcoming barriers to ensuring the right balance of practical relevance and intellectual rigour in research.

Doing so is important as, until fairly recently, there had been comparatively little attention paid to this debate or its implications within CMR (Bresnen, 2017). This is despite the amount of critical debate in business and management circles (Grey, 2001; Bresnen and Burrell, 2013) and despite the self-evident importance of combining rigour and relevance in *any* form of management research. The core idea of 'engaged scholarship' (Van de Ven, 2007) has certainly seen some direct application to CMR (Voordijk and Adriaanse, 2016), and these and related knowledge production concepts and ideas have figured in several conference contributions (Bresnen, 2001; Fernie and Leiringer, 2009; Harty and Leiringer, 2007, 2008). However, it is only in the last few years that this debate has moved centre stage into CMR via the pages of its leading journals and, in doing so, has started to infuse wider discourse and debate within the field on the nature and future of CMR itself.

Promoting and provoking that discussion, in 2017 the editors of *Construction Management and Economics* launched a landmark special issue that called for more discussion and debate on the future of CMR (Leiringer and Dainty, 2017a, 2017b). This prompted serious questioning about the future trajectory of CMR in theoretical and empirical terms (Harty and Leiringer, 2017). It also raised important questions about the practical relevance of management research (Koskela, 2017) and the challenges of translating theory and research from one management domain to another to promote rigour and relevance (Bresnen, 2017). Building upon a long-standing interest within CMR in promoting methodological rigour (Seymour and Rooke, 1995; Raftery et al., 1997) and theoretical strength and diversity (Schweber, 2015), the special issue brought questions of rigour and relevance into sharp focus, sparking lively debate (e.g. Ivory, 2017) and heralding a renewed emphasis on achieving rigour and relevance through research and publication (Chan, 2020). Subsequent notable contributions to the debate (Ivory, 2017; Koch et al., 2019; Volker, 2019) have reasserted the long-established view that social science has a good deal to offer the study of construction management phenomena (Bresnen et al., 2005). Yet, they have also continued to focus on issues in the translation of management research into CMR (Fellows and Liu, 2020) and to question whether enough is yet being done to mobilise insights from cognate fields to generate intellectually rigorous and practically useful research (Dainty and Leiringer, 2019; Chan, 2020).

While there is no question that existing research within CMR is directly informed by (a diversity of) theoretical perspectives and that contributions are driven ultimately by highly salient industry needs, this is not quite the same thing as ensuring rigour and relevance, as many editorials focusing on the review process are keen to point out (e.g. Javernick-Will, 2018; Chan, 2020). As

Schweber (2015, p. 840) has noted, CMR as an academic field is centred upon a domain of practice, rather than being embedded in a bedrock discipline. As such, and like BMR, it combines the advantages of being able to draw upon a wide range of disciplinary perspectives, with the disadvantages that come with theoretical eclecticism. Unlike BMR, however, for the most part it has not nurtured within it the core concepts and frameworks that underpin management research. Instead, these have been largely imported or 'borrowed' from outside the field (Bresnen, 2017; Fellows and Liu, 2020). What this means for CMR, according to Schweber (2015, p. 841), is the need for a deeper engagement with theories at their source rather than their selective (and superficial) application. This, in turn, requires reflexivity in theoretical positioning and the approach to research. Elsewhere, I have asked the question of how CMR can benefit from its engagement with BMR, and what the implications are of greater (or lesser) institutional cross-over between academic fields (Bresnen, 2017, p. 25). Here, my intention is to explore these themes further, with particular attention directed to some of the inherent tensions and contradictions in the relationship between rigour and relevance in BMR. The central message should emerge that, while such an aim is highly laudable and very desirable, its achievement remains a challenging one.

Rigour and relevance in BMR

The rigour–relevance debate has been a prominent feature of BMR since the mid-1990s (e.g. Gibbons et al., 1994; Pettigrew, 1995) and has surfaced the tensions that can exist between the academic value of the outputs that stem from it (i.e. publications) and the impact of that work on policy and practice. A number of prominent interventions through keynote presentations at the American academy (Huff, 2000; Huff and Huff, 2001; Bartunek, 2007; Shapiro, 2016) helped spark debate about the balance between the intrinsic academic value and the practical impact of management research. The benefits of promoting more 'engaged scholarship' (Van de Ven, 2007) were proposed as a way of re-balancing research to ensure that it retained its practical relevance, which many feel it had lost in the search for academic rigour and scientific credibility.

Similar trends were apparent in the UK, where a report on management research policy, instigated by the British Academy of Management, recommended more practically focused, trans-disciplinary research in order to avoid what it saw as the pitfalls of "epistemic drift" and "academic fundamentalism" (Tranfield and Starkey, 1998, p. 353). The report and its recommendations later formed the centrepiece of a special issue of the *British Journal of Management*

(Hodgkinson, 2001). This attracted a number of contributions – both supportive and critical – that sought to address the value of more practically oriented, trans-disciplinary research (Starkey and Madan, 2001; Grey, 2001; Huff and Huff, 2001; Pettigrew, 2001; see also Hodgkinson and Rousseau, 2009; Starkey et al., 2009).

This debate centred upon a core idea that had developed within the science policy field that what we have been witnessing is an inevitable (and desirable) historical transition from one mode of knowledge production (Mode 1) to another (Mode 2). While the former privileged traditional scientific values and academic institutions, the latter stressed the importance of trans-disciplinary research, co-produced in the context of application (Gibbons et al., 1994; Nowotny et al., 2001, 2003). In *The New Production of Knowledge*, it was argued that the conditions shaping knowledge production in advanced industrial economies were shifting massively and profoundly affecting how knowledge was being produced and applied in ways that questioned the hegemony of traditional scientific disciplines (Gibbons et al., 1994, p. 13). The result was a paradigm shift in the approach to research and an emphasis upon research that had five distinct key features (Nowotny et al., 2001).

The first feature is *knowledge produced in the context of application*, with research driven by the need to address practical problems and articulated by a variety of stakeholders to meet 'market needs'. This contrasts with earlier Mode 1 research, where problems were supposedly wholly defined (and solutions produced) by academic and professional communities working within and through their specialist scientific disciplines. In Mode 2, the needs of business and other key stakeholders would be more important in defining the goals of research. Nowotny et al. (2001) described this as 'contextualised science' and likened the process of mobilising research to the *agora* in ancient Athens – a public arena for the open regulation of science, where scientists, citizens, and consumers could meet and effectively transact to meet their research needs. Second, *transdisciplinarity* becomes the basic approach to research. This goes beyond assembling a diverse range of specialists to work in teams on problems and, instead, involves them developing a common framework, shaped in the context of application, to guide practical, problem-oriented research – using a range of theoretical perspectives and research methods that may not easily reduce to core disciplinary scientific knowledge or epistemic practices. Third, greater *institutional heterogeneity* means that research contributions stem from a wider range of organisations than hitherto (including research and consultancy organisations, government agencies, and companies), whose involvement (and even existence) may be flexible and temporary, but which together signifies a radical shift in the institutional locus of research. Fourth, research becomes *reflexive and dialogical*, involving 'an intense (and perhaps endless) "conversation" between research actors and research subjects' (Nowotny et al., 2001, p. 187), allowing multiple views and perspectives to shape research processes and outcomes, thus lessening deference to 'objective' scientific methods. Finally, *social accountability* is emphasised, with scientific peer review being superseded by new forms of quality control that incorporate a wider range of criteria (economic, social, political) for judging the outcomes of research.

While the central tenets of this call for more co-produced, applied research have persisted over time (Hodgkinson and Rousseau, 2009), contributors have suggested a number of refinements to the model. Huff (2000), for example, initially suggested a Mode 1.5 as a more disciplinary grounded approach that reflected the reality faced by academics in otherwise having to compete with others (i.e. consultants) who are 'skilled at seeking Mode 2 work and know how to perpetuate it' (Huff, 2000, p. 291). Rather than being a midway position, Mode 1.5 involves a continued valorisation of academic research in business schools, but with academics having to 'improve their ability to convey the importance of the work they do' (Huff, 2000, p. 290). Reflecting, too, the importance of a wider range of societal interests than businesses per se, Huff and Huff (2001) proposed a Mode 3 option, which responded directly to the need to 'modernise' research, but which took a more communitarian line in emphasising a wider range of stakeholders than just business and the value of seeing knowledge as a public good.

The development of this thesis, in turn, helped shape continuing wider debate within the UK and American academy about the balance between academic rigour and practical relevance, leading to a succession of journal special issues in the *Academy of Management Journal* (2007), *Journal of Management Studies* (2009), and *Organisation Studies* (2010). Indeed, Hessels and van Lente (2008, p. 749) identified over 1000 journal citations up to 2007, most of which appeared in the introductions and conclusions of articles, suggesting a high degree of infusion in academic discourse. More pertinently, perhaps, the approach has become a taken-for-granted feature of the research funding landscape, as reflected in the importance of targeted funding initiatives, 'grand challenges', and the 'impact agenda' in research commissioning and assessment (Thorpe et al., 2011). Consequently, while it originated as a treatise on the changing nature of science practice with implications for science policy, Mode 2 and its variants have since become a manifesto for the conduct of BMR (including, by implication, CMR).

Modes of knowledge production: issues and debates

While the approach has been influential both directly and indirectly through its impact on discourse on BMR, it has not been without its criticisms which, according to Hessels and van Lente (2008, pp. 755–6) fall into three main categories: lack of empirical support, conceptual incoherence, and normative underpinnings.

Regarding empirical support, the original thesis contained little case evidence to substantiate its claims about a full-scale transition to Mode 2. Although there may be a general sense in which heterogeneity in research and pressures towards greater social accountability are increasing, many commentators have either questioned the extent to which these trends are occurring, or suggested that they have always figured in research in the applied and social sciences (Godin, 1998; Godin and Gingras, 2000; Weingart, 1997; Ziman, 1996). The same commentators point to even less evidence of change in other Mode 2 characteristics, suggesting that the changes occurring are more limited in scope, application, and impact than the thesis initially presumed. Even advocates admit there were over-inflated claims made about the scale of change being experienced (Starkey and Madan, 2001) and admit to 'infatuation' with the idea (Hodgkinson and Starkey, 2011, p. 360). While journal citation or content analysis gives some indication of the scale, breadth, and depth of the impact of the debate (Hessels and van Lente, 2008; Nicolai and Siedl, 2010), examples of truly reflexive 'research on research' are hard to find. Where Mode 2 is explicitly the object of study, researchers have pointed to the challenges of reaching consensus on the aims, perspectives, and methods of collaborative research (Mitev and Venters, 2009) and have also questioned the value of knowledge sometimes produced (MacLean et al., 2002, p. 203).

Regarding *conceptual coherence*, many continue to be highly critical of the distinction drawn between Mode 1 and Mode 2. Ziman (1996, 2000), for instance, questions the conceptual clarity, distinctiveness, and novelty of Mode 2, arguing that traditional (Mode 1) science placed a lot of emphasis on research being problem driven, and that creative and innovative research did occur, albeit often at the margins rather than at the core of scientific disciplines. He also questions the presumption in Mode 1 of personal disinterestedness, given the pressures within professional scientific communities that steered research towards 'acceptable' lines of enquiry. It is also questionable whether 'post-academic science' gives scientists any real freedom to cross discipline boundaries, given the continued importance of disciplinary career structures. The effort required to deal with highly structured mechanisms governing grant applications also makes the 'attractively unbureaucratic' nature of post-academic science a myth (Ziman, 2000, pp. 80–1). Others similarly contest the idea of a simple and wholesale transition from Mode 1 to Mode 2 research in the applied and social sciences (Etkowitz and Leydesdorff, 1999; Godin, 1998; Rip, 2002; Weingart, 1997).

While some accept that a distinct and coherent form of research can and does exist (McLean et al., 2002), others are more sceptical, questioning the internal lack of coherence of the concept and its weak or absent theoretical underpinnings (Rip, 2002; Shinn, 2002). Clearly, the categorisation of all forms of science into two main types is simplistic (Rip, 2002), as even the thesis's proponents have been obliged to recognise (Nowotny et al., 2001; Starkey and Madan, 2001). Even where alternatives are explored (e.g. Hodgkinson et al., 2001), it is hard not to see in these efforts a clear preference for one type of science over others – or, until recently, any attempt to transcend the associated dualism of rigour and relevance (Hodgkinson and Starkey, 2011; Thorpe et al., 2011).

Regarding Mode 2's *normative underpinning*, that tendency is apparent in the conflation of description, analysis, and prescription that characterises the original thesis, despite claims made to the contrary:

> No judgement is made as to the value of these trends – that is, whether they are good and to be encouraged, or bad and resisted – *but it does appear that they occur most frequently in those areas which currently define the frontier and among those who are regarded as leaders in their various fields.* (Gibbons et al., 1994, p. 1, emphasis added)

Indeed, it has been argued that, in subordinating academic science to market forces, Mode 2 simply represents the replacement of one power/knowledge configuration with another and, in the process, reinforces and legitimises any consequent loss of academic autonomy (Grey, 2001). Moreover, as Ziman (2000) notes, the proprietary control of research can inhibit the development and spread of knowledge. It also has implications for what is taken for granted as either 'useful' or 'useless' research (Learmonth et al., 2012), an implied threat made manifest by Hodgkinson et al.'s (2001) categorisation of some forms of research as 'pedantic' or 'puerile'. Such categorisations are not only judgemental and dismissive. They also take a static view of the relevance of research that fails to account not just for shifts in relations of power that determine research agendas (Grey, 2001; Willmott, 2012), but also paradigm shifts that determine the locus of relevant research within disciplines. Moreover, they fail to take into account the susceptibility of (business and management) research to changing managerial fads and fashions (Abrahamson, 1996).

Commentators have also been quick to point out how these criticisms feed also into the wider rigour–relevance debate. Despite the extensive support that does exist for the idea that rigour and relevance can and should be combined (Hodgkinson, 2001; Pettigrew, 2001; Bartunek, 2007; Hodgkinson and Rousseau, 2009), there are many who emphasise the challenges and trade-offs in trying to do so (e.g. Gulati, 2007; Rynes et al., 2007; Beech et al., 2010). This is due to major differences in academic and practitioner aims, perspectives, and methods (Pettigrew, 2001), which some even suggest creates a fundamental incompatibility of academic and practitioner approaches – with relevance often only being achieved at the expense of rigour and vice versa (Kieser and Leiner, 2009). Beech et al. (2010) explore in some depth the different embedded assumptions that militate against productive dialogue. They argue that, instead of the more relational view of knowledge co-production emphasised in some accounts that promotes active and continuing dialogue based on closer emotional engagement (Bartunek, 2007), the best that can be hoped for are generative dialogic encounters that attempt to overcome disparate goals and means in academic–practitioner interaction.

Even the notion of 'relevance' has been shown to be ill-defined (Hodgkinson, 2001) and to have multi-faceted instrumental, conceptual, and legitimising meanings underpinning it (Nicolai and Seidl, 2010). The same, of course, applies to the conception of 'rigour', if we take into account the diverse epistemic practices across disciplines that affect the generation of scientific knowledge more generally (Knorr-Cetina, 1999), as well as the production of management knowledge more specifically (Sahlin-Andersson and Engwall, 2002). The challenge then becomes one of developing a deeper understanding of how theory and practice link together in the design and conduct of research (Jarzabkowski et al., 2010).

Institutional enablers and impediments to flows of management knowledge

Given this wider debate, the question also emerges of the challenges and opportunities faced by CMR, given the gap which exists between that community and those with interests in BMR. Elsewhere, I have suggested that the two domains can be seen as distinct communities of practice (Bresnen, 2016, 2017), with their own perspectives and distinct ways of knowing (Boland and Tenkasi, 1995). They also occupy very different institutional terrains, with quite separate structures and processes in place for the development, furtherance, realisation, and assessment of research (Bresnen, 2017; Harty and Leiringer, 2017). In such a context, management knowledge may act as a 'boundary object' (Star and Griesemer, 1989) whose 'plasticity' helps bridge the gap between these distinct institutional domains (Bresnen, 2017). As a boundary object, management knowledge might be plastic enough to enable conversations to occur and knowledge to flow across this divide. However, the danger is that the way it is constituted might make it too *elastic* and prone to create further barriers to the effective translation of that knowledge across disciplinary boundaries (Fellows and Liu, 2020).

The challenge for CMR is in overcoming what could be described as a 'triple hurdle' of engaging with a very disparate and often hotly contested management knowledge base, while overcoming the 'double hurdles' (Pettigrew, 1995) of reconciling research rigour and relevance (Bresnen, 2017, p. 30). Taking the double hurdles first, the challenge is reflected in many recent calls from leading figures in the field to ensure that a general concern for the practical import of CMR is actually realised, while at the same time ensuring that research contributions are sufficiently rigorous (Koskela, 2017; Leiringer and Dainty, 2017a; Dainty and Leiringer, 2019; Chan, 2020). CMR has always taken its lead from a concern with industry issues. However, recent concerns have been expressed that not only is there a danger of losing sight of the practical application of research (Koskela, 2017); there is also the continuing need to attend to the academic rigour which underpins CMR research (Dainty and Leiringer, 2019; Chan, 2020).

While the first point is highly debatable (Ivory, 2017), a good illustration of the second point can be found in research that has taken as its focus the challenge of understanding and researching temporary organisations (Sydow et al., 2004; Bakker et al., 2016). One of the defining features of project-based construction work is its temporal finality, and that has been long realised as key to understanding organisation and management in construction (e.g. Bryman et al., 1987). Yet, in-depth examination of the consequences of temporal finality for understanding construction project organisation has, for the most part, been left to organisational theorists (outside CMR) interested in the fundamental nature of project-based organisation and/or in the 'projectification' of contemporary organisational forms (Lundin and Söderholm, 1995, 1998; Sydow et al., 2004). Consequently, it is in the realm of BMR, rather than CMR, that we tend to find closer interrogation of the ontology of project-based organisations (including construction), as well as of the epistemological complexities of researching them (e.g. Bakker et al., 2016). This is not to suggest that BMR is better suited to that task or more complete in that respect. For example, the application of an alternative 'becoming' ontology (Tsoukas and Chia, 2002) that helps us understand better the rhythms of project work and the fragility of project organisation built upon them is still in its infancy (Bakker et al., 2016). Nor is it meant to suggest that construction project organisations do not figure in those debates. Indeed, there have been many important contributions (e.g. Söderlund, 2011; Van Marrewijk et al., 2016). However, it does suggest that those fundamental debates are largely happening elsewhere and that, in CMR, they remain, at best, under-explored issues and, at worst, largely side-stepped or taken for granted.

Harty and Leiringer (2017) have gone so far as to suggest that CMR faces four possible future trajectories: first, *convergence*, around a core (and narrow) set of topics, problems, methods, and theoretical perspectives; second, *retrenchment*, through a return to a more practically oriented research and teaching agenda; third, *disappearance*, as a separate community, through incorporation into other institutional structures (business and engineering schools); and fourth, *hybridisation*, in which the strengths of CMR are retained but in ways that require closer connectivity and engagement with wider networks and institutional structures. The authors present the latter as a more radical option in which the rigour–relevance debate is transcended through a clearer engagement of CMR with wider BMR and impact agendas, with a corresponding need for radical change to institutional structures within the CMR field.

Underpinning this debate, however, is the challenge thrown up by the wide diversity of BMR that, in turn, makes the generation and accumulation of knowledge through CMR itself highly challenging. With regard to overcoming this third hurdle, I have pointed out elsewhere (Bresnen, 2017) that management knowledge is highly fragmented, situated, mediated, provisional, and contested (Blackler, 1995). There are several reasons for this. First, management itself comprises multiple 'communities of knowing' (Boland and Tenkasi, 1995) associated with distinct subject areas of interest (e.g. human resources, finance, marketing, operations). Specialisation itself is not at issue, but it does mean that management knowledge is itself highly distributed and occupies a rather sublime position at the interstices between disciplines in this multi-disciplinary subject space. Moreover, each of these subject areas has, in turn, its own distinct research agendas and questions and its own predilections for different forms of research (more theoretically or practically defined; more quantitative or qualitative). The inevitable outcome is a plurality of aims, perspectives, and approaches that might cohere around key research questions, 'grand challenges', or explicit multi-disciplinary projects, but which inevitably brings together researchers with distinct epistemic practices or ways of knowing (Knorr-Cetina, 1999; Schweber, 2015). The result is perhaps what could best be described (in inter-disciplinary research) as a *collectivity*, rather than community, of knowledge and practice (Lindkvist, 2005). Consequently, the challenge is already how to select from, focus on, and mobilise management knowledge to meet the diverse research agendas these subject areas represent.

Second, management knowledge is itself the product of a complex amalgam of academic and industry knowledge that has been shaped over time in different ways and to varying degrees by practitioners, both within companies or public sector bodies or at one remove by third parties (i.e. consultants). Again, distinct discipline areas have been influenced differently in the extent to which research exhibits features more redolent of Mode 1 or Mode 2. Not only does this mean that management knowledge bears the imprint of those involved in its (co-)production. It also means that it is borne directly out of the circumstances of its social construction. Management accounting knowledge, for example, betrays its origins in the role of the US military (Hoskin and Macve, 1998); and 'leadership' is a word of Anglo-Saxon origin for which there is no direct equivalent in the French language. Moreover, the translation of management knowledge into everyday practice is often associated with a transformation of that knowledge as management perspectives and preferences shape their embedding in organisational practices (Sahlin-Andersson and Engwall, 2002).

Third, like many other areas of knowledge, BMR is shaped by the influence of changing business and other (e.g. government) preferences for funding research that inevitably favour different forms of knowledge according to what are seen as the business and societal challenges of the time. What this presents is a highly fluid and somewhat contested and politicised notion of what constitutes 'relevance'. Moreover, unlike many other areas of knowledge, BMR is arguably more affected by the ebb and flow of management fads and fashions (Abrahamson, 1996). Notable developments in management thought (scientific management, culture management, business process reengineering, the knowledge economy, and the like) have all had distinct phases of influence and attention in both management and academic circles (e.g. Grint, 1994). They also bear the hallmark of commodification, given the commercial or public value of the knowledge created and the role of consultants in the generation, mobilisation, and translation of such knowledge (Sturdy, 2011; Engwall and Kipping, 2013).

Last but not least, management knowledge, particularly in some sub-disciplines, is also hotly contested (Clegg and Palmer, 1996), with multiple theories and major differences between opposing schools of thought continuing to provoke ontological and epistemological debates between different paradigmatic viewpoints on the same subject area (Burrell and Morgan, 1979). As a common object, management knowledge is therefore not only subject to different readings from different perspectives, it is also constituted in complex and constantly changing ways (Ivory, 2017). The challenge is then to be able to draw carefully upon such a diverse and potentially incompatible range of ideas, by ensuring appropriate contextualisation of their theoretical and empirical origins, as well as their conceptual and methodological strengths and weaknesses (Sage et al., 2014; Fellows and Liu, 2020).

This challenge is added to by the observation that conceptions of rigour and relevance are themselves constituted differently, insofar as they proceed from very different ontological and epistemological presumptions (e.g. about the value to different stakeholders of management knowledge; Grey, 2001; Willmott, 2012; Bresnen and Burrell, 2013). In contrast to the more pessimistic view that BMR has become too distanced from practical relevance (Koskela, 2017), those within critical management studies in BMR would argue instead that research in BMR has privileged performativity and the primacy of managerial interests for too long, leading to a major imbalance in management research (Fournier and Grey, 2003; Alvesson and Willmott, 2000). This does not mean that the search for practical relevance is unimportant or should lessen or discontinue. However, it does suggest a wider conception of the notion of relevance that meets multiple stakeholder interests (Huff and Huff, 2001) and an opportunity to draw upon a range of alternative perspectives (both mainstream and critical) to inform and shape the generative dialogue associated with collaborative research (Beech et al., 2010).

Implications for CMR

What does all this mean for CMR? What are the implications for how researchers might deal with the challenges of these triple hurdles of engaging in rigorous and relevant research while at the same time navigating this complex and changing terrain?

Building on the suggestions of recent contributors to the debate (Schweber, 2015; Ivory, 2017; Fellows and Liu, 2020), one critical way forward is the need for researchers to apply a more reflective and reflexive approach that fully takes into account the situated nature of BMR management knowledge, its conceptual and methodological foundations and fault-lines, and its nuances and fragilities. BMR knowledge is diverse, complex, changeable, and contested, and recognising that is the first step in avoiding selectivity and uncritical acceptance of particular theories, models, and concepts. It also helps prevent too much deference to hegemonic and/or fashionable paradigms, perspectives, or approaches. A good example of this is the emergence of more critical perspectives on projects and project management (Hodgson and Cicmil, 2006) which has helped promote greater dialogue around the aims, objectives, and methods of project management research, while still promoting rigour and relevance.

Indeed, elsewhere, it has been argued that there is greater scope now than ever before for CMR to shape more directly BMR and so avoid too much of what has historically been a one-way flow of knowledge between BMR and CMR (Bresnen, 2017). As noted earlier, important developments have emerged in recent years in understanding temporary organisation (Bakker et al., 2016) and the 'projectification' of work and organisation (Lundin and Söderholm, 1998; Sydow et al., 2004). However, the same is true of work on supply chain governance (Clegg et al., 2002), the integrating power of design objects (Ewenstein and Whyte, 2009), and changing forms of employment (Dainty et al., 2007). Research in these areas (and others) is helping to shape and inform wider debates about the changing nature of organisation and management in late modernity (see also Sage et al., 2016). Developments like these offer the opportunity for greater connectivity and cross-fertilisation of ideas between CMR and BMR in ways that promote the hybridisation suggested by Harty and Leiringer (2017).

However, it is not just more judicious attention to the underlying management knowledge base that is important but, as Harty and Leiringer (2017) suggest, what is important too is change in research practices and the surrounding institutional conditions that shape them. Recognising that institutional continuity or change can be shaped at different levels depending on the agency of individuals or groups acting individually or collectively (Lawrence et al., 2011), what is suggested here is that there are three levels at which attention might be focused and challenges faced: the institutional, the relational, and the individual.

First, at the *institutional level*, I have suggested elsewhere (Bresnen, 2017) that there is a good deal of work required at the institutional level (Lawrence et al., 2011), particularly if the challenge of change involves making CMR more hybrid (Harty and Leiringer, 2017), with research cutting across established institutional structures promoting research and publication. While institutional researchers have long struggled with the challenge that embedded agency poses to agents of change (Battilana and D'Aunno, 2009), some have emphasised too the opportunities to shape institutional conditions through institutional work (Lawrence et al., 2011) and to exploit internal contradictions through *praxis* (Seo and Creed, 2002). *Praxis* is the artful mobilisation of action by individuals and groups to exploit internal institutional contradictions that can arise in four principal ways (Seo and Creed, 2002, p. 226): privileging legitimacy (that undermines functional efficiency); generating adaptations (that inhibit adaptability); fostering internal conformity (that creates inter-institutional incompatibilities); and promoting isomorphism (that clashes with divergent interests). It is not too difficult to see in these internal contradictions the opportunity they allow for key institutional actors to shape the nature and direction of CMR by highlighting internal contradictions in how research is governed and mobilised. This can be through, for example, promoting and shaping debates about future research directions in leading journals, on funding body panels, at major international/national conferences, in key practitioner–academic forums, and through university- and faculty-level decisions on research funding, assessment, and support.

Second, at a *relational level*, the argument could be made that there is more need for researchers to complement the development of strong bonds within the CMR community of practice (as valuable as that is) with greater bridging social capital that promotes a more outward-looking approach (Putnam, 1995; Burt, 1997). Bridging social capital builds on the strength of weak ties, opening up any community of practice to a wider range of knowledge and experience and encourages divergence of viewpoints and openness, rather than tendencies towards conformity and network closure (Uzzi, 1997). While institutional change promises to alter conditions at the source through the manipulation of internal institutional contradictions, there is still the need for active engagement by those within the community of practice with those outside. Practical steps might involve the exploration of and engagement with wider networks of research (and teaching) within BMR (and beyond) that speak to CMR research interests but would normally be beyond the institutional reach of that community of practice.

Third, at an *individual level*, there is the concomitant need for researchers to engage reflexively with their own roles and identities as construction management researchers. Creating more of a two-way exchange of knowledge and ideas between CMR and BMR challenges researchers in both fields to engage more fully with the wider set of interests and networks in which the other operates (see Sage, Chapter 12, this volume). In turn, academic institutions can do a lot to either positively or negatively reinforce these tendencies, through decisions that are made in assessing research outputs and promotion applications, in supporting research applications and conference attendance, and in providing research training and support (Bresnen, 2017, p. 31).

Returning to the earlier discussion of the shift in modes of knowledge production, there is a final irony here in the heuristic value that such an approach potentially has in helping CMR overcome the challenging triple hurdles of combining rigour and relevance with effective engagement with BMR given the complex, multi-layered nature of this challenge. It should be clear from the earlier review that the 'Mode 2 thesis' has been found wanting in a number of ways in terms of its adequacy in describing, analysing, and explaining current developments in BMR. Yet, despite these limitations, it has undoubtedly had an impact on BMR research policy and practice and the discourse surrounding that, and it continues to underpin important debates about ensuring rigour and relevance in research. Echoing some of the points made above about the potential value of seemingly less relevant or 'useful' research in both BMR (Learmonth et al., 2012) and CMR (e.g. Ivory, 2017), there is an ironic twist here in that the normative underpinnings of the framework itself nevertheless do at least provide a useful heuristic for assessing the developmental journey of CMR. In continuing to ensure that CMR combines rigour and relevance in appropriate measure (Dainty and Leiringer, 2019; Chan, 2020) and does so by engaging as appropriate with BMR (Sage et al., 2014; Fellows and Liu, 2020), there is arguably some value to be had in assessing evolving conditions against that framework. It does, after all, promote knowledge produced in the context of application, trans-disciplinarity, institutional heterogeneity, reflexivity/dialogue, and new forms of quality control. Arguably, that is the direction of travel flagged up by some (e.g. Harty and Leiringer, 2017) as a potential route forward for ensuring CMR continues to combine rigour with relevance. At the very least, it signals that even the most controversial and criticised ideas in BMR still have some traction.

Indeed, applying this logic suggests some very practical ways in which the community can respond to the challenges noted earlier. Given that institutional-level change can be difficult to engineer in the short-to-medium term and that institutional contexts also constrain individual action (though reinforcing disciplinary norms and career structures), the more obvious starting point is at the relational level. Given the centrality of project-based organising noted earlier, it is perhaps apposite to focus on *the project* as the vehicle for change. In developing new research projects and in building the network of participants around them, there are clearly significant opportunities for embracing change. The *context of application* for research can be as much related to the interests driving forward research in BMR as to home-grown agendas based on extant CMR knowledge and industry needs. Opportunities for *transdisciplinarity* may be enhanced further, with project (and writing) teams explicitly comprised of both CMR and BMR researchers (as well as industrial partners and/or other key stakeholders). *Institutional heterogeneity* will be reflected not only in the institutional affiliation of participants, but also in the diversity of forms of outlet sought for research outputs (including conferences papers, workshops, and teaching programmes). Provided there is sufficient reflexivity (including not treating 'extraneous' knowledge as a project 'add in'), the *generative dialogue* developed through project-based interaction (and associated wider networks) may help foster an approach that helps infuse problem-focused CMR research with wider knowledge and interests, allowing it to speak to wider issues and agendas. *New forms of quality control* can in turn reflect researchers' own self-efficacy in developing publication and impact plans.

These suggestions are not meant to lessen the importance of continuing to promote change at an institutional level to create appropriate enabling conditions. That is still critically important in the medium-to-long term. But they do present practical starting points to help ensure the greater agility needed to rise to the challenges of hybridisation and avoid other, arguably much less attractive, future options and trajectories for CMR.

References

Abrahamson, E. (1996). Management fashion. *Academy of Management Review, 21*(1), 254–85.

Alvesson, M., and Willmott, H. (2003). *Studying management critically*. London: SAGE.

Bakker, R. M., DeFillippi, R. J., Schwab, A., and Sydow, J. (2016). Temporary organising: Promises, processes, problems. *Organisation Studies*, 37(12), 1703–19.

Bartunek, J. M. (2007). Academic–practitioner collaboration need not require joint or relevant research: Toward a relational scholarship of integration. *Academy of Management Journal, 50*(6), 1323–33.

Battilana, J., and D'Aunno, T. (2009). Institutional work and the paradox of embedded agency. In T. B. Lawrence, R. Suddaby, and B. Leca (Eds), *Institutional work: actors and agency in institutional studies of organisations*, pp. 31–58. Cambridge: Cambridge University Press.

Beech, N., MacIntosh, R., and MacLean, D. (2010). Dialogues between academics and practitioners: The role of generative dialogic encounters. *Organization Studies, 31*(9–10), 1341–67.

Blackler, F. (1995). Knowledge, knowledge work and organizations: An overview and interpretation. *Organization Studies, 16*(6), 1021–46.

Boland, R., and Tenkasi, R. (1995). Perspective making and perspective taking in communities of knowing. *Organization Science, 6*(4), 350–72.

Bresnen, M. (2001). The new production of knowledge: Implications and challenges for construction management research? Keynote Presentation in A. Akintoye (Ed.), *Proceedings*, ARCOM 17th Annual Conference, University of Salford, 5–7 September 2001.

Bresnen, M. (2016). Institutional development, divergence and change in the discipline of project management. *International Journal of Project Management, 34*, 328–38.

Bresnen, M. (2017). Being careful what we wish for? Challenges and opportunities afforded through engagement with business and management research. *Construction Management and Economics*, *35*(1–2), 24–34.

Bresnen, M., and Burrell, G. (2013). Journals à la mode? Twenty years of living alongside Mode 2 and the new production of knowledge. *Organisation*, *20*(1), 25–37.

Bresnen, M., Goussevskaia, A., and Swan, J. (2005). Managing projects as complex social settings. *Building Research and Information*, *33*(6), 487–93.

Bryman, A., Bresnen, M., Beardsworth, A., Ford, J., and Keil, E. (1987). The concept of the temporary system: The case of the construction project. In S. Bacharach (Ed.), *Research in the Sociology of Organisations*, Vol. 5, pp. 253–84. London: JAI Press.

Burrell, G., and Morgan, G. (1979). *Sociological paradigms and organizational analysis*. London: Heinemann.

Burt, R. S. (1997). The contingent value of social capital. *Administrative Science Quarterly*, *42*(2), 339–66.

Carlile, P. (2004). Transferring, translating, and transforming: An integrative framework for managing knowledge across boundaries. *Organization Science, 15*(5), 555–68.

Chan, P. W. (2020). Revisiting basics: Theoretically-grounded interesting research that addresses challenges that matter. *Construction Management and Economics*, *38*(1), 1–10.

Clegg, S., and Palmer, G. (1996). *The politics of management knowledge*. London: Sage Publications.

Clegg, S., Pitsis, T. S., Rura-Polley, T., and Marroszeky, M. (2002). Governmentality matters: Designing an alliance culture of inter-organizational collaboration for managing projects. *Organization Studies*, *23*(3), 317–37.

Dainty, A., Green, S., and Bagilhole, B. (2007). People and culture in construction: Contexts and challenges. In A. Dainty, S. Green, and B. Bagilhole (Eds.), *People and culture in construction: A reader*, pp. 3–25. London: Spon.

Dainty, A., and Leiringer, R. (2019). Maintaining a relevant construction management and economics research community. *Construction Management and Economics*, *37*(12), 693–6.

Engwall, L., and Kipping, M. (2013). Management consulting: Dynamics, debates, and directions. *International Journal of Strategic Communication, 7*(2), 84–98.

Etkowitz, H., and Leydesdorff, L. (1999). The future location of research and technology transfer. *Journal of Technology Transfer*, *24*, 111–23.

Ewenstein, B., and Whyte, J. (2009). Knowledge practices in design: The role of visual representations as 'epistemic objects'. *Organization Studies*, *30*(1), 7–30.

Fellows, R., and Liu, A. M. M. (2020). Borrowing theories: Contextual and empirical considerations. *Construction Management and Economics*, *38*(7), 581–8.

Fernie, S., and Leiringer, R. (2009). The construction management field: a bit of a dig. In *Proceedings of the 5th Nordic Conference on Construction Economics and Organisation*, vol. 1, Reykjavik, Iceland, June 10–12, pp. 110–20.

Fournier, V., and Grey, C. (2000). At the critical moment: Conditions and prospects for critical management studies. *Human Relations, 53*(1), 7–32.

Gibbons, M., Limoges, C., Nowotny, H., Schwartzman, S., Scott, P., and Trow, M. (1994). *The new production of knowledge: The dynamics of science and research in contemporary societies*. London: SAGE.

Godin, B. (1998). Writing performative history: The new New Atlantis. *Social Studies of Science*, *28*, 465–83.

Godin, B., and Gingras, Y. (2000). The place of universities in the system of knowledge production. *Research Policy, 29*(2), 273–78.

Grey, C. (2001). Re-imagining relevance: A response to Starkey and Madan. *British Journal of Management, 12*(SI), S27–S32.

Grint, K. (1994). Reengineering history: Social resonances and business process reengineering. *Organisation, 1*(1), 179–201.

Gulati, R. (2007). Tent poles, tribalism and boundary spanning: The rigour–relevance debate in management research. *Academy of Management Journal, 50*, 775–82.

Harty, C., and Leiringer, R. (2007). Social science research and construction: Balancing rigour and relevance. In *Proceedings of the Construction Management and Economics 25th Anniversary Conference*, Reading, UK, July 16–18.

Harty, C., and Leiringer, R. (2008). Construction management researchers as agents of change: Balancing academic priorities and industry impact. In *Proceedings of the CIB W55/65 Joint International Symposium*, Dubai, November 15–17.

Harty, C., and Leiringer, R. (2017). The futures of construction management research. *Construction Management and Economics, 35*(7), 392–403.

Hessels, L. K., and van Lente, H. (2008). Re-thinking new knowledge production: A literature review and a research agenda. *Research Policy, 37*, 740–60.

Hodgkinson, G. P. (Ed.). (2001). Facing the future: The nature and purpose of management research re-assessed. *British Journal of Management, 12*(SI), S1–S80.

Hodgkinson, G. P., Herriott, P., and Andersen, N. (2001). Re-aligning the stakeholders in management research: Lessons from industrial, work and organizational psychology. *British Journal of Management, 12*(SI), S41–8.

Hodgkinson, G. P., and Rousseau, D. M. (2009). Bridging the relevance gap in management research: It's already happening! *Journal of Management Studies, 46*, 534–46.

Hodgkinson, G. P., and Starkey, K. (2011). Not simply returning to the same answer over and over again: Reframing relevance. *British Journal of Management*, 22, 355–69.

Hodgson, D., and Cicmil, S. (Eds). (2006). *Making projects critical.* Basingstoke, UK: Palgrave Macmillan.

Hoskin, K. W., and Macve, R. H. (1998). The genesis of accountability: The West Point connections. *Accounting, Organisations and Society, 13*(1), 37–73.

Huff, A. S. (2000). Changes in organisational knowledge production. *Academy of Management Review, 25*(2), 288–93.

Huff, A. S., and Huff, J. O. (2001). Re-focusing the business school agenda. *British Journal of Management, 12*, S49–S54.

Ivory, C. (2017). The prospects for a production management body of knowledge in business schools: Response to Koskela (2017) "Why is management research irrelevant?" *Construction Management and Economics, 35*(7), 385–91.

Jarzabkowski, P., Mohrman, S. A., and Scherer, A. G. (2010). Organization studies as applied science: The generation and use of academic knowledge about organizations. *Organization Studies, 31*(9–10), 1189–1207.

Javernick-Will, A. (2018). Rationale: The necessary ingredient for contributions to theory and practice. *Construction Management and Economics, 36*(8), 423–4.

Kieser, A., and Leiner, L. (2009). Why the rigour–relevance gap in management research is unbridgeable. *Journal of Management Studies, 46*, 516–33.

Knorr-Cetina, K. (1999). *Epistemic cultures. How the sciences make knowledge.* Cambridge, MA: Harvard University Press.

Koch, C., Paavola, S., and Buhl, H. (2019). Social science and construction – An uneasy and underused relation. *Construction Management and Economics, 37*(6), 309–16.

Koskela, L. (2017). Why is management research irrelevant? *Construction Management and Economics*, *35*(1–2), 4–23.
Lawrence, T., Suddaby, R., and Leca, B. (2011). Institutional work: Refocusing institutional studies of organization. *Journal of Management Inquiry*, *20*(1), 52–8.
Learmouth, M., Lockett, A., and Dowd, K. (2012). Promoting scholarship that matters: The uselessness of useful research and the usefulness of useless research. *British Journal of Management*, *23*, 35–44.
Leiringer, R., and Dainty, A. (2017a). Construction management and economics: New directions. *Construction Management and Economics*, *35*(1–2), 1–3.
Leiringer, R., and Dainty, A. (2017b). On the right to disagree, healthy debates and moving forward. *Construction Management and Economics*, *35*(7), 383–4.
Lindkvist, L. (2005). Knowledge communities and knowledge collectivities: A typology of knowledge work in groups. *Journal of Management Studies*, *42*(6), 1189–1210.
Lundin, R. A., and Söderholm, A. (1995). A theory of the temporary organization. *Scandinavian Journal of Management*, *11*, 437–55.
Lundin, R. A., and Söderholm, A. (1998). Conceptualizing a projectified society: Discussion of an eco-institutional approach to a theory on temporary organizations. In R. A. Lundin and C. Midler (Eds.), *Projects as arenas for renewal and learning processes*, pp. 13–23. Norwell: Kluwer.
MacLean, D., MacIntosh, R., and Grant, S. (2002). Mode 2 Management Research. *British Journal of Management*, *13*, 189–207.
Mitev, N., and Venters, W. (2009). Reflexive evaluation of an academic-industry research collaboration: Can mode 2 management research be achieved? *Journal of Management Studies*, *46*(5), 733–54.
Nicolai, A., and Seidl, D. (2010). That's relevant! Different forms of practical relevance in management science. *Organization Studies*, *31*(9–10), 1257–85.
Nowotny, H., Scott, P., and Gibbons, M. (2001). *Re-thinking science: Knowledge and the public in an age of uncertainty*. Cambridge: Polity.
Nowotny, H., Scott, P., and Gibbons, M. (2003). 'Mode 2' revisited: The new production of knowledge. *Minerva*, *41*(3), 179–94.
Pettigrew, A. M. (1995). *The double hurdles for management research*. Distinguished Scholar Address to the OMT Division of the US Academy of Management, Vancouver, Canada, August.
Pettigrew, A. M. (2001). Management research after modernism. *British Journal of Management*, *12*, S61–S70.
Putnam, R. D. (1995). Bowling alone: America's declining social capital. *Journal of Democracy*, *6*, 65–78.
Raftery, J., McGeorge, D., and Walters, M. (1997). Breaking up methodological monopolies: A multi-paradigm approach to construction management research. *Construction Management and Economics*, *15*(3), 291–7.
Rip, A. (2002). Regional innovation systems and the advent of strategic science. *Journal of Technology Transfer*, *27*(1), 123–31.
Rynes, S. L. (2007). Editor's forward – Carrying Sumantra Ghoshal's torch: creating more positive, relevant and ecologically valid research. *Academy of Management Journal*, 50, 4, 745–747.
Rynes, S. L., Giluk, T. L., and Brown, K. C. (2007). The very separate world of academic and practitioner publications in human resource management: Implications for evidence-based management. *Academy of Management Journal*, *50*, 987–1008.

Sage, D., Dainty, A., and Brookes, N. (2014). A critical argument in favour of theoretical pluralism: Project failure and the many and varied limitations of project management. *International Journal of Project Management, 32*(4), 544–55.
Sage, D., Justesen, L., Dainty, A., Tryggestad, K., and Mouritsen, J. (2016). Organizing space and time through relational human–animal boundary work: Exclusion, invitation and disturbance. *Organization, 23*(3), 434–50.
Sahlin-Andersson, K., and Engwall, L. (2002). Carriers, flows and sources of management knowledge. In K. Sahlin-Andersson and L. Engwall (Eds.), *The expansion of management knowledge*, pp. 3–32. Stanford, CA: Stanford Business Books.
Schweber, L. (2015). Putting theory to work: The use of theory in construction research. *Construction Management and Economics, 33*(10), 840–60.
Seo, M. G., and Creed, W. E. D. (2002). Institutional contradictions, praxis and institutional change: A dialectical perspective. *Academy of Management Review, 27*, 222–47.
Seymour, D., and Rooke, J. (1995). The culture of the industry and the culture of research. *Construction Management and Economics, 13*(6), 511–23.
Shapiro, D. L. (2016). Presidential address. Delivered at *Making organizations meaningful*, the 76th Annual Meeting of the Academy of Management, Anaheim, California, August 7.
Shapiro, D. L., Kirkman, B. L., and Courtney, H. G. (2007). Perceived causes and solutions of the translation problem in management research. *Academy of Management Journal, 50*(2), 249–66.
Shinn, T. (2002). The triple helix and new production of knowledge: Prepackaged thinking on science and technology. *Social Studies of Science, 32*, 599–614.
Söderlund, J. (2011). Pluralism in project management: Navigating the crossroads of specialization and fragmentation. *International Journal of Management Reviews, 13*, 153–76.
Star, S. L., and Griesemer, J. R. (1989). Institutional ecology, 'translations' and boundary objects: Amateurs and professionals in Berkeley's Museum of Vertebrate Zoology, 1907–39. *Social Studies of Science, 19*, 387–420.
Starkey, K., Hatchuel, A., and Tempest, S. (2009). Management research and the new logics of discovery and engagement. *Journal of Management Studies, 46*, 547–58.
Starkey, K., and Madan, P. (2001). Bridging the relevance gap: Aligning stakeholders in the future of management research. *British Journal of Management*, 12, S3–S26.
Sturdy, A. (2011). Consultancy's consequences? A critical assessment of management consultancy's impact on management. *British Journal of Management, 22*(3), 517–30.
Sydow, J., Lindkvist, L., and DeFillippi, R. (2004). Project-based organisations, embeddedness and repositories of knowledge. *Organization Studies, 25*(9), 1475–89.
Thorpe, R., Eden, C., Bessant, J., and Ellwood, P. (2011). Rigour, relevance and reward: Introducing the knowledge translation value chain. *British Journal of Management, 22*, 420–31.
Tranfield, D., and Starkey, K. (1998). The nature, social organization and promotion of management research: Towards policy. *British Journal of Management, 9*, 341–53.
Tsoukas, H., and Chia, R. (2002). On organizational becoming: Rethinking organizational change. *Organization Science, 13*(5), 567–82.
Uzzi, B. (1997). Social structure and competition in interfirm networks: The paradox of embeddedness. *Administrative Science Quarterly, 42*, 35–67.
Van de Ven, A. H. (2007). *Engaged scholarship: A guide for organisational and social research*. Oxford: Oxford University Press.

Van de Ven, A. H., and Jonson, P. E. (2006). Knowledge for theory and practice. *Academy of Management Review*, *31*(4), 802–21.
Van Marrewijk, A. H., Ybema, S., Smits, K., Clegg, S., and Pitsis, T. S. (2016). Clash of the Titans: Temporal organizing and collaborative dynamics in the Panama Canal Megaproject. *Organization Studies*, *37*, 1745–69.
Volker, L. (2019). Looking out to look in: Inspiration from social sciences for construction management research. *Construction Management and Economics*, *37*(1), 13–23.
Voordijk, H., and Adriaanse, A. (2016). Engaged scholarship in construction management research: The adoption of information and communications technology in construction projects. *Construction Management and Economics*, *34*(7–8), 536–51.
Weingart, P. (1997). From 'finalization' to 'Mode 2': Old wine in new bottles? *Social Science Information*, *36*(4), 591–613.
Willmott, H. (2012). Reframing relevance as 'social usefulness': A comment on Hodgkinson and Starkey's 'Not simply returning to the same answer over and over again'. *British Journal of Management*, *23*, 598–604.
Ziman, J. M. (1996). Postacademic science: Constructing knowledge with networks and norms. *Science Studies*, *9*(1), 67–80.
Ziman, J. M. (2000). *Real science: What it is and what it means.* Cambridge: Cambridge University Press.

4 Theory and the contribution of qualitative research to construction management research

Libby Schweber and Vivien Chow

Introduction

The title of this chapter points to a problem that the authors see in qualitative studies in construction management research (CMR). In many cases, researchers and PhD students eschew theory development in a supposedly modest attempt to hew to the empirical data and to speak to industry, making descriptive claims that are limited to their observations. Having offered a thematic description of what they 'observed', authors then use their analysis to make very general (immodest) claims about what practitioners in all situations, at all times, should do. All too often those claims come in the form of factors that practitioners should pay attention to or barriers they should overcome or yet another model of the relations between discrete variables, thereby offering an illustration of what Cornelissen (2017) refers to as 'quantitative restyling' (p. 368). This approach involves a reduction of qualitative research to a quantitative, positivist model and, in doing so, severely limits its contribution. At the heart of this problem is a misunderstanding of the role of theory in interpretivist or qualitative research and an attempt to superimpose a positivist, engineering approach. The purpose of this chapter is not to document this critique – which, like many characterisations, captures something about the current state of the art, but does not apply to every case. Instead, it is to make a case for the heuristic use of theory in qualitative research and to argue for the importance of this type of theorising for both academic and practical contributions. Theory, we argue, is our main and perhaps only tool for moving beyond received views and professional common sense and, hence, a condition sine qua non, for making a contribution.

This chapter is one of those relatively rare pieces in academic research – a commissioned text – where the authors are given a topic and asked to write about it. In our case, the remit was to reflect on the contribution of qualitative research. When we started, we assumed that we would write about the ongoing debate on rigour vs relevance (see below). However, in the process of reviewing secondary literature on the contribution of academic research and analysing a collection of what we deemed to be good qualitative research papers in CMR, we came to the view that concern over this tension is misplaced, at least when it comes to CMR. Stated differently, we find that the solution to both greater rigour and greater relevance is the same: namely, a more rigorous use of theory and more systematic engagement with theory development. Given the common view that theory is what makes our papers inaccessible to practitioners, this position clearly needs to be explained. This chapter attempts to do that. Before starting, however, it is perhaps helpful to lay out our initial positions and to clarify what we do and do not mean.

We start from three assumptions. These are: 1) like management research, CMR is positioned at the intersection of multiple knowledge communities (Lindkvist, 2005); 2) CMR should address industry and practical problems, but it should do so in novel ways that challenge received ways of thinking (common sense); and 3) theory is the primary tool that we, as researchers, have to accomplish this. Knowledge communities include academia, industry, policy makers, and others with a stake in the built environment. Each of these communities involves distinctive ways of thinking and organising. The contribution of CMR lies in bringing our distinctive way of thinking to these different communities, rather than in effacing it. While academics should not compromise their way of thinking, they should connect to other knowledge communities.

Our second assumption starts from the conventional and, for some, contested position (e.g. Kilduff and Kelemen, 2001) that academics can best reach different audiences by selecting problems that are relevant for them (Jarzabkowski et al., 2010), pursuing them 'scientifically' by mobilising theory to explore those problems, and communicating the findings through multiple, diverse types of outputs targeted at distinct audiences. In taking this as a starting point, we do not dismiss calls for practitioner involvement in the production of research (Bartunek, 2007; Van de Ven, 2007, 2018), but rather complement it by drawing attention to what academic researchers bring to the table, independent of whether they are involved in 'pure' research, co-production (Jasanoff, 2004; Antonacopoulou, 2009), Mode 2 (Gibbons et al., 1994), or some other form of industry–academic engagement.

Finally, our third assumption is that one, if not the, main difference between academic knowledge and other types lies in the use of theory. Theory, we argue, is what allows scholars to introduce novel perspectives, to disrupt dominant taken-for-granted understandings, and to come up with novel and interesting perspectives that open new possibilities. Without it, academic research is no different from consultancy or industry and policy reports or journalism – all valuable types of knowledge in their own right, but often without the potential to move beyond conventional understandings.

One corollary of this position is our view that academic journals are the appropriate place for explicit discussion and reflection of theory, that they have a constitutive role in the reported research and contributions to theory development, and that their audience is other academics, not practitioners. Stated more bluntly, industry and policy makers do not and should not be expected to read academic journal publications (MacIntosh et al., 2017; Bartunek, 2007). Viewed from this perspective, it is surprising how much of the debate on rigour vs relevance combs academic publications in search of practical relevance, instead of considering the variety of types of outputs from a given research project, such as industry and policy reports, consultancy work, and other forms of industry training (Nicolai and Seidl, 2010).

In sum, our argument in this chapter is that the production of research should be epistemologically rigorous and theoretically informed; that academic journals are the appropriate place to demonstrate rigour; that collectively we should aspire to theory development; and that theory development will enhance our contribution to ongoing academic conversations, within CMR and beyond, and to practice. The point about contributing to both CMR and other disciplines relates to a discussion in the CMR academic literature about whether theoretical work in CMR should focus on the development of a distinctive core of CMR theory or whether we should be borrowing from and contributing to other more mature disciplines (Fellows and Liu, 2020; Leiringer and Dainty, 2017a, 2017b). Here, too, our argument is that this is a false choice. The same type of work, we contend, is essential for both.

The discussion that follows explores these points. We start with a brief discussion of the specificity of qualitative research. We introduce the idea that theory development – and, by extension, the 'contribution' of qualitative research – resides in the analysis of mechanisms and how they play out in different contexts. Having established a baseline for the discussion, we reflect on tensions informing qualitative research, including the ongoing discussion of rigour vs relevance. We then set out our argument that theory and theory development are key to knowledge contribution in qualitative research. Following this, we turn to the concept of 'theory' itself. Building on Sandberg and Alvesson's (2021) distinction between five types of theory, we exemplify each type using examples from the qualitative CMR literature. The aim of the discussion is both to map out a range of types of theoretical contribution and to consider what is meant by a heuristic (rather than representational) use of theory. Our review provides an opportunity to reflect on the current use of theory in qualitative CMR papers and their contribution. The chapter concludes with recommendations for how CMR scholars and gatekeepers of the discipline can enhance the contribution of our work to both academics and practitioners by engaging more systematically and rigorously with theory and theory development.

Qualitative research

The term 'qualitative research' is generally used to refer to research with qualitative data and to a specifically qualitative approach or epistemology. Qualitative data can be analysed from a positivist or quantitative approach, or it can be analysed from an interpretivist or qualitative approach (for a discussion of this distinction applied to CMR, see Schweber, 2014). Following the convention adopted in this collection, we use the term qualitative to refer specifically to interpretivist approaches.

Research approaches can be characterised by the type of ontology that they adopt (their assumptions about 'reality'), by the type of questions that they ask, by the type of theories that they use and develop, and by the type of claims that they seek to make. Historically, positivist social science can be traced to the work of Auguste Comte and his view that social life can be studied in the same way as the natural world. Fast-forward to the 20th century, and this basic premise supports a rationalist approach in which theoretically derived laws are translated into hypotheses or propositions about the relations between variables or distinct elements in a model and tested with empirical data. The type of knowledge produced has been described as 'representationalist' (Tsoukas, 1998) or 'knowledge by representation' (Chia and Holt, 2008). These terms point to the assumption that positivist models accurately represent 'reality', if not perfectly, then well enough to support practical intervention and prediction. Implicit in this approach is an assumption of relatively constant relations which can be observed, such that all instances of a phenomenon will evidence the same patterns. While such claims are qualified by scope conditions, there is an aspiration to identify general laws.

Among positivist researchers, quantitative data is often held up as the gold standard. That said, the approach can be applied to qualitative data, including to the analysis of rich interview data. Within qualitative research in CMR, the dominance of this approach can be seen in the analysis of rich interviews for key barriers or factors deemed to 'explain' a particular outcome. The study of factors or barriers is positivist in the sense that it explores the relation between relatively fixed discrete factors or elements and their effect on a discrete outcome. It is worth noting that 'explanation' in this approach relies on probabilistic relationships, not on how causal factors produced the observed outcome. Theory in this work is often reduced to a descriptive model or system, which is seen to apply similarly across instances (for a discussion of this approach within business school knowledge practices, see Jarzabkowski et al., 2010).

Qualitative or interpretivist approaches almost always begin from a critique of this quantitative or positivist approach. In the late 19th and early 20th centuries, intellectual developments, including hermeneutics, neo-Kantian philosophy, and phenomenology, all challenged the premise that social life could be studied in the same way as natural life. At the heart of the critique was an assumption that people differed from physical objects in the role of meaning and symbols in shaping individual behaviour, social action, and the course of history. Within the social sciences, these trends led to a focus on *verstehen*, or interpretation of and by social actors. In contemporary research, this critique has developed into a general approach that starts from an assumption that social phenomena are shaped by meaning-laden processes.

Whereas quantitative approaches ask 'what'-type questions about patterns of relationships between distinct variables, qualitative approaches ask 'how'- and 'why'-type questions about holistic clusters of elements. Qualitative researchers are interested in how a particular phenomenon came to assume one form rather than another; the processes which explain a particular outcome; and variations in the same phenomenon across actors, settings, and time. In recent discussions of management research, this approach leads to calls for scholars to pay attention to context and time (Pettigrew and Starkey, 2016).

The underlying theoretical question in most qualitative research (even if it is not stated) is how to account for either continuity or change. This point is important, as it suggests that qualitative researchers must explain why things stay the same as well as how and why they change, and to do so they must inquire into the meanings that different actors ascribe to those things and the processes by which they were produced. Whereas quantitative researchers often ask about the relations between variables in fixed states, qualitative researchers study processes. For example, in the hands of interpretivists, the study of strategy becomes a study of (situated) strategising and the study of innovation becomes a study of (situated) innovating. Instead of probabilistic relations, explanation is usually about mechanisms that work out differently in different contexts, and the focus is on variations rather than similarities across instances or cases.

The concept of mechanism is central to theorising in qualitative research and, as such, deserves a bit more explanation. In a classic article, Stinchcombe (1991) defines 'mechanisms' as 'bits of theory about entities at a different level (e.g. individuals) than the main entities being theorised about (e.g. groups), which serve to make the higher-level theory more supple, more accurate, or more general' (p. 367). Starting from this definition, theory development around mechanisms depends on: 1) specifying a broader theory, 2) identifying a theoretical phenomenon to explore through that lens, 3) selecting an empirical case that exemplifies that problem and allows for its investigation, 4) developing a theoretically informed research design to investigate the way in which a particular process (which makes up the selected phenomenon) plays out in that specific (empirical setting), and 5) theorising from that instance (or those instances) around the type of mechanisms that produced that outcome.

A couple of comments help to flush out this research process. First, as illustrated in the discussion to follow, the choice of theory informs the empirical research question. Second, the empirical study of a case is always also a study of a specific type of context. For example, in the case of project-level research in CMR, it often involves an inquiry into how a particular phenomenon plays out in a complex, project-based, multi-disciplinary, multi-firm, temporary organisation. One of the main contributions of CMR qualitative research to both the development of discipline-specific knowledge and to other disciplines lies in theorisation around the specificity of that context and its effect on already theorised processes, be it strategy implementation, innovation, or gender discrimination, to name but a few.

Finally, very few scholars contribute entirely new theories, using either quantitative or qualitative research. Looking at the past 50 years or so, the number of genuinely novel theories can probably be counted on one or two hands. Instead, the primary contribution of most qualitative research and, especially, of work informed by industry-based problems lies in the development of middle-range theory (Merton, 1967, Ch. 2; Green and Schweber, 2008; Birkinshaw et al., 2011; Makadok et al., 2018) and specifically in the identification and nuancing of mechanisms.

A key point regarding theorisation is that while one case can be used to identify or propose a particular mechanism (for theory development), one case does not a theory make. Instead, theory development in qualitative research is necessarily a collective activity in which successive studies, by the same or other scholars, are used to explore the way in which a particular mechanism works out in different contexts. It is, by its nature, a comparative process. While more post-modern theories, such as action-network theory (ANT), may reject this type of theorisation a priori, we suggest that the contribution to knowledge – be it about goal setting or about power or about innovation – loosely fits this kind of reasoning. The discussion that follows offers a number of illustrations.

Tensions in academic research

Academic research in general, and management research and CMR more particularly, are fraught with tensions. Researchers need to navigate the tension between the specificity of empirical data and generalisation or theory; between pressure to publish and passionate or risky research (Courpasson, 2013); between different stakeholders and audiences; and, according to an ongoing debate within the literature, between relevance and rigour. The debate over relevance and rigour is an internal academic debate that has been going on for more than 40 years (MacIntosh et al., 2012) and has received considerable attention among gatekeepers of the discipline, notably editors and keynote speakers at management conferences. Pettigrew (2001) famously referred to rigour and relevance as the 'double hurdles' through which management research must jump. The debate takes different forms ranging from concern over research training, to the output and function of business schools (Chia and Holt, 2008; Aram and Salipante, 2003), to reflections on the relation between science and industry (as in the debate over Mode 2 knowledge), to more classical discussions of the relation between theory and practice (Jarzabkowski et al., 2010; Van de Ven, 2007). Within the UK, a version of the argument has informed funding councils, as evidenced in their growing demands to reflect on and evidence impact and practitioner engagement. Within management studies, observers are divided as to whether this is a central, constitutive issue which every researcher and academia as a whole must address, or a largely symbolic meta-discussion with little to no impact on scholarly practice (MacIntosh et al., 2012; Carton and Philippe, 2017). Mike Bresnen's contribution to this book makes a case for the import of the relevance–rigour debate for CMR. In this section, we briefly visit this debate, with an aim of indicating how it does and does not apply to qualitative research.

The rigour–relevance debate usually focuses on a positivist version of research, associated with business schools and management studies. The critique is that the rationalist or representationalist type of knowledge produced is far from actionable, leading to a tension between the requirements of academic research and industry practitioners. The representationalist approach aims at capturing the true principles and features of the industry so that firms can make rational, informed choices to get their desired results (Chia and Holt, 2008). The more rigorously positivist strictures for research are followed, the greater the accuracy of the findings and relevance for practitioners – or so representationalists claim. For some scholars, the practical irrelevance of this approach can be ascribed to the focus on macro- rather than micro-level analysis or on general acontextual rather than situation-specific knowledge (Aram and Salipante, 2003). For others, it stems from an excessive concern with formal knowledge at the expense of personal experience (Van de Ven, 2007) or tacit (Chia and Holt, 2008) or colloquial knowledge (Kondrat, 1995). For yet others, the problem lies in methodological rigour, which comes at the expense of genuine insight (Aguinis et al., 2014). This 'academic rigour' position is contrasted with the interest of industry in context-specific solutions to immediate short-term problems.

When it comes to qualitative research, many of the key points in this debate seem misplaced or irrelevant. Where quantitative research aims for universal generalizations and formal knowledge, divorced from practitioner experience, qualitative research focuses on local cases and the effect of context. When it comes to practitioners, the aim is to offer them a new perspective on their own experiences and understandings.

That said, where the debate does seem relevant is in the contribution of qualitative research. The practical contribution of qualitative academic research does not lie in reproducing practitioner views or in solving short-term problems. Instead, it lies in an exploration of the multiple interpretations that different actors and stakeholders bring to a situation and in the way in which those differences and the actions which ensue contribute to understandings of continuity or change. As Pierre Bourdieu has argued, the contribution of qualitative research lies in the displacement of their current understandings by even a small amount, and the suggestions of new, alternate, interesting, and provocative ways of thinking about what they already think about (Bourdieu et al., 1991; Bourdieu and Wacquant, 1992).

A similar point can be made about the contribution of qualitative research to academic research. The question of what makes research 'interesting' has been the topic of extensive discussion, notably in editorial comments. Curiously, it echoes Bourdieu's call for research that displaces readers' common sense. For example, Bansal and Corley (2011) underline the ability to provide readers with an 'Aha' moment, in which rich (and we would add, theoretically informed) accounts allow readers to suddenly see or understand something which they did not previously. Similarly, Davis (1971) argues that topics are interesting when they counter readers' taken-for-granted assumptions. In a parallel argument, Alvesson and Sandberg (2013) critique the deadening effect of 'gap spotting' in management studies and argue for innovating and influential research which challenges underlying assumptions.

For all of these authors, theory and theory development are the researcher's primary tool to move beyond taken-for-granted views. Whereas representationalist approaches focus on the fit between models and data, interpretivist researchers (ideally) search for places where 'encounters between theoretical assumptions and empirical impressions' (Alvesson and Sandberg, 2013, p. 146) break down, as it is there that interesting contributions originate and theory development begins. As this suggests, the contribution of qualitative research lies not in telling academics what they already know or practitioners what to do, but in helping them to think differently, moving them beyond the trap of received views and dominant understandings, and offering novel and provocative ways of thinking about what they think they already know.

On the matter of theory

Even the most superficial review of the literature indicates that there are numerous definitions of 'theory' (Colquitt and Zapata-Phelan, 2007). More positivist definitions focus on a positivist type of causal explanation (e.g. Whetten, 1989). More interpretivist definitions treat theory (or bits of theory) as tools to explore messy and complex phenomena through the lens of a particular set of constructs. We refer to this as a heuristic use of theory. In one such definition, Suddaby (2014) writes that 'theory is simply a way of imposing conceptual order on the empirical complexity of the phenomenal world' (p. 407).

A key problem in qualitative CMR is the limited understanding and use of theory. Many scholars adopt a positivist version of theory. Unable to see the relevance of this type of analysis, they abandon the endeavour completely, offering their qualitative research as a unique description of an individual case. In a recent article, Sandberg and Alvesson (2021) argue for the existence of multiple types of theory, with different purposes. Their paper defines 'theory' in terms of six criteria, leading them to identify five different types. In terms of this essay, these typologies offer a way to think about theory and to expand the range of what is considered. We introduce them here and use them to describe five CMR papers in the section that follows.

The six criteria include: 1) having a purpose, 2) being directed at a phenomenon, 3) offering some form of conceptual order, 4) providing intellectual insights, 5) specifying relevance criteria, and 6) having empirical support. Readers will recognise a number of these points from the discussion above. Viewed from the perspective of this essay, Sandberg and Alvesson's discussion offers a preliminary answer to the question: How do qualitative researchers make a contribution? They use a set of abstract concepts (a theory) to explore an empirical issue or problem by treating it as an instance of a broader phenomenon, exploring key features of the empirical instance, and, in doing so, breaking with common-sense understandings to offer an interesting and 'useful' account for either academic or practical purposes. The square brackets in the following sections signal a reference to one of the six criteria.

The five purposes or types of theory are: explaining, ordering, comprehending, enacting, and provoking. As the authors note, the types are not mutually exclusive. Instead, papers may have more than one purpose (Sandberg and Alvesson, 2021, pp. 495–7). The discussion which follows uses five CMR empirical research papers to illustrate these types. The aim is to expand the dominant CMR understanding of what counts as theory and thereby as a contribution. The papers were selected for their ability to illustrate key points – some are exemplary, others offer excellent starting points, but could be developed further either by the author or in future research.

Explaining

Explaining theory takes for its purpose a causal account of a phenomenon. This type of explanation dominates positivist research. In this representationalist version of explaining, theory involves describing a phenomenon in terms of discrete variables, exploring the causal (probabilistic) relations between them, and offering a rationale for why they are related in that way (explaining). Within interpretivist research, most empirical papers have an 'explaining' purpose, in the sense that they set out to solve an empirical puzzle, but this is almost always accompanied by one of the other four purposes that inform the way they represent the problem and the type of explanation they proffer.

Within CMR qualitative literature, a key limitation of many supposedly 'explaining' papers is reduction of theory to a model (a representationalist approach). While models are essential components of explaining theories (they list the different factors or entities that are seen to be related), they do not in and of themselves constitute a theory. Instead, theory includes claims concerning the relations between the elements in the model and their effects (or in this case, explanation). A second common limitation, especially in qualitative research, is the tendency to limit claims to a single case study, without engaging in theory development at all. The result is a kind of false objectivism that purports to accurately describe what 'actually' happened, denying both their own selectivity in producing the account and adding little to current understandings. One example of an explaining paper that uses a single empirical case but avoids these limitations is Sminia's (2011) account of institutional continuity in the construction sector. The paper starts with a clear empirical question and offers a theoretically informed response.

The paper begins with a puzzle. The author asks: Why did the Dutch construction industry continue to engage in pre-consultation for nine years after the EU ruled it illegal? The phenomenon is not pre-consultation (the empirical case) but rather regulation and compliance in the industry, and even more broadly, institutional continuity [phenomenon]. The purpose is, first and foremost, explanatory – to answer the empirical question. In addition, the explanation offered involves the deployment of an enacting theory, namely, neo-institutionalism (NI).

As a theory, NI draws attention to the way in which culturally embedded practices reproduce existing activities. It also highlights the impact of institutional tensions on the enactment of those practices. The findings highlight the existence of a cluster of practices, including particular ways of tendering, pre-bidding, and contracting. The main theoretical argument concerns the ways in which these practices are shaped by institutional tensions between 1) limited budgets versus elaborate wants, 2) price versus value, and 3) unlawful but fair choices versus lawful but unfair choices. In terms of theory development, a key contribution lies in the classification of types of tensions and in the analysis of the dynamic generated. It is worth underlining the distinctiveness of these three tensions or analytical constructs. The distinction between (un)lawful and (un)fair choices is original and does not simply reproduce professional common sense. This originality gives the paper an 'aha' moment which moves the reader beyond usual policy maker or senior management explanations [intellectual insights]. The three tensions identified are specific to the empirical case, but they can also be expected to figure in other cases; the theoretical question is how far to generalise. While a single study cannot conclusively establish scope conditions [relevance], it can speculate, and fellow researchers can further explore the relevance of the theory for other types of cases. For example, the paper explains how the failure to recognise and address structural tensions inherent in construction work (and previously managed by pre-consultation activities) undermines efforts at change. A similar problem can be seen in attempts to address the housing shortage in the UK, where the failure to address underlying problems with dominant business models and financial arrangements undermine government attempts to stimulate building.

Ordering

A review of theoretically informed CMR articles suggests that ordering is one of the least common types of theorising. The purpose of ordering theory is to categorise a phenomenon into multiple types, thereby identifying variations and contributing to theory development. Classic management examples of ordering theories include Minzberg's (1979) distinction between five types of firm structures and Porter's (1980) typology of firm strategies. Each of these well-known typologies has been used to explore a wide range of different empirical problems, while successive empirical studies have been used to refine and nuance the theory. The same cannot be said of ordering theories within CMR.

Three tendencies would seem to account for the paucity of ordering theorising. The first is the tendency of qualitative researchers to focus on single cases, making comparison across types less likely. The second involves the representationalist approach of many qualitative researchers, which leads them to treat typologies as descriptions. Instead of theorising the dynamics of each type and the conditions accounting for variations in their expression, most CMR papers use the typology to order their data. The third involves the lack of cumulative research agendas within CMR (Schweber, 2014), and means that few scholars pick up on new typologies and explore their use in the explanation of different but related empirical problems.

An example of ordering theory in the CMR literature which introduces a promising set of analytic distinctions but stops short of theorising can be found in Chi and Javernik-Will's (2011) comparative study of high-speed railway projects in China and Taiwan. The [purpose] of the paper is to explore the effect of national-level institutions on the delivery of large, innovative infrastructure projects. Contained in this formulation is the authors' choice of institutional theory, which directs attention to the role of cultural as well as organisational institutions [conceptual order]. Specifically, the paper explores the effect of national-level political culture and industry structure on project delivery [phenomenon]. In this paper, the effect of differences in the institutional context was found to have been exercised through a number of mechanisms, including: the location of decision-making authority, the place of the project in national political strategies, and firm size and capabilities [intellectual insights].

The identification of these analytic constructs offers the beginning of a theoretical and practical contribution, but the paper stops short of theorising these mechanisms. The authors present their practical contribution as offering contextual considerations for foreign firms considering bidding for large infrastructure projects; however, their insights are limited to the two empirical cases. What is missing is a more systematic reflection on the different locations where decision-making takes place in the construction sector more generally, on variations in the place of large infrastructure projects in national strategies, on the effect of firm size and capabilities on project governance, and on the effect of different combinations of these mechanisms. For example, one would want to know whether a case with centralised decision-making authority and a national political strategy which took construction for granted would generate similar or different dynamics to the Chinese case documented in the paper. Similarly missing is any reflection on scope conditions, leaving the reader wondering on whether these different types of construction regime are only relevant for high-speed rail projects, or whether their influence might extend to all types of large infrastructure projects or to construction more generally. Without this, the reader is left with a persuasive description, but no propositions or more abstract theoretical claims for further exploration in other cases. When it comes to ordering theory, the contribution lies not in the introduction of new analytic distinctions, but in a systematic reflection on the types of relationships and processes which different configurations in different settings might be expected to produce and an associated research agenda for further exploration.

Comprehending

Comprehending theories often start from the role of language in social life. A key premise is that meaning shapes and even produces organisations. Examples of comprehending theories include discourse analysis, narrative analysis, and social identity theory. In all three cases, the focus is on how peoples' interpretations and understandings are structured, how those understandings interact, and how that dynamic produces the organisational phenomenon of interest. One of the limitations of CMR research is the tendency to limit analysis to what senior managers say, without considering either the 'hidden meanings' beneath their words or alternate perspectives. Within CMR, there is a tendency to impose a representationalist approach onto this type of analysis by treating the theory as a proposition to be tested. This leads to studies which include in their findings a 'validation' of the theory – as in the unsurprising claims that construction professionals tell stories or have professional identities. This claim involves a misunderstanding of both positivist and interpretivist research. In positivist research it is an empirical proposition, derived from a theory, which is tested – not an entire theory. In interpretivist research, theories are tools to explore rich empirical data; the places where the empirical predictions do not fit (not the descriptions that 'fit') are of the most interest and open the way to theory development.

Löwstedt and Räisänen's (2014) paper offers an example of a theoretically informed comprehending paper that goes beyond face-value claims by senior managers. The purpose of their paper is to analyse the way in which identity work – defined as the way in which individuals draw on existing narratives to construct their sense of who they are and what they do – influences organisations [purpose]. Like Sminia (see above) the authors are interested in the study of continuity, rather than change. Starting with the case of a large engineering and consultancy firm, with multiple hierarchical roles and both geographic and functional divisions, the authors ask: How do the employees identify, and how does this mode of identification affect the implementation of top-down strategic change initiatives [phenomenon]? [Conceptual ordering] figures in the authors' analytic distinction between in-groups and out-groups and the way in which that boundary figures in identity work among their research subjects. Interesting findings include employees' strong identification as construction workers (rather than firm or division members, for example) and as short-term problem solvers or doers who do not engage with change visions [intellectual insights].

As in Sminia and Chi and Javernik-Will's papers, the 'aha' moment, empirical contribution, and potential for theory development all stem from the introduction of theoretically informed analytic constructs (in this case, boundaries between in- and out-groups) to analyse otherwise familiar situations. The finding that construction workers identify as members of an occupation, rather than a firm, challenges the usual focus on firm branding and firm-level identity, offering a surprising insight and helping to explain the failure of employees to engage with strategic change initiatives. Rather than seeing the construction industry as conservative and slow to change, we might instead begin to understand how a strong group identity allows workers to navigate complex and volatile construction site environments. Moving beyond the confines of this individual paper, the analysis also introduces a new research agenda focused on the scope of these findings. Is this type of primary identification unique to certain types of construction firms? To certain countries? To certain sectors? To particular roles? And how does it vary? As this example illustrates, the theoretical contribution of the paper does not lie in simply describing the case studies through the lens of these analytic distinctions, but rather in an analysis of the dynamics generated around them in relation to a particular phenomenon (in this case, continuity of practice and rejection of management fixes). As in the previous examples, theory development is a collective affair, calling on fellow researchers to pick up and run with theoretical suggestions.

Enacting

Enacting theory is one of the most common types of theorising in qualitative CMR research. This can partly be attributed to the recent 'popularity' of a number of enacting theories among qualitative CMR researchers, including ANT, social practice theory, and activity theory. At their core, enacting theories posit a dynamic ontology in which phenomena are studied as processes, rather than fixed states. Within CMR, many enacting theories start from the failure of most managerial and technological fixes to deliver as promised. Instead of assuming a linear, seamless delivery process, enacting theorists seek to account for organisational processes. A limitation in many CMR enacting papers is their focus on a particular empirical case rather than a phenomenon. The result is a descriptive, contingent account of how a particular instance or case unfolded, without any contribution to either theory development or practice.

Koch and Schultz (2019) offer an example of an enacting paper. The [purpose] of the paper is to explore the ongoing production of defects in construction projects. As this formulation indicates, the paper begins from a social constructivist ontology. Defects are treated as the result of having been named and treated as such, rather than intrinsic features of the building. This is, or should be, a key feature of all qualitative interpretivist papers; all too often CMR researchers slip into treating their phenomenon as given, thereby adopting an essentialist, positivist ontology. The focus of the paper is thus not on defects, but the problem-solving process by which defects are produced and handled [phenomenon]. To explore this phenomenon, the authors use a version of structuration theory.

The theory focuses attention on the dynamic interaction between structures and agency. One difficulty of this theory is the very high level of abstraction of the key constructs, which makes it difficult to engage directly with the data and to make a genuine contribution. Koch and Schultz deal with this by focusing on a couple of key conceptual distinctions. The most fruitful, to our minds, is their use of the distinction between three types of agency: reactive, proactive, and here-and-now types of agency [conceptual order]. The interest of this analytic distinction lies in the 'aha' moments it produces. A key finding is the dominance of reactive agency – a point which is at odds with arguments about the pragmatic, here-and-now focus construction activity more generally (e.g. Daudigeos, 2013). More generally, the paper shows how defects are not an anomaly, but a result of established problem-solving practices [intellectual insights].

The authors suggest that the practical contribution of the study stems from following the project over an extended period. While this type of longitudinal research design is to be applauded, it does not involve the type of theoretical work required to apply a meso-level theory to the micro-level dynamics of everyday project work. One of the difficulties in this paper is the gap between the very abstract level of theory – focusing on types of structure and agency – and the empirical data and an associated failure to theorise problem solving in construction (as opposed to problem solving in general). As in the other papers, the specification of scope conditions is a crucial first step to theory development, opening the way for a nuancing of the theoretical contribution. This would seem to be particularly important for CMR scholars, where a key collective contribution is a systematic reflection on the specificity of construction work and the way in which broad abstract analytic distinctions play out in construction-specific contexts. Future theorising could explore the type of problem solving at play in standard versus bespoke projects (a distinction which the authors themselves introduce but do not systematically analyse), as well as an exploration of variations across other types of construction settings.

Provoking

Provoking theory takes the imperative for novel or theoretical approaches the furthest. The main purpose of this type is to present an alternative, and even disruptive, vision of how contemporary organisation might be radically different. The challenge is both to how things are and to how they are understood to be, with some authors assuming that the two are one and the same. Whereas in all of the types of theory examined thus far, the purpose is (or should be) to reconstruct a phenomenon differently, in provoking theory this aim is explicitly directed at challenging and changing power relations.

A common move in provoking theorising is a critical analysis of dominant metaphors and the introduction of a different one, with all of the possibilities that this reconceptualisation provides (for a discussion of the metaphors informing dominant CMR discourse, see Green et al., 2008). Of the five types, provoking theory is the least common in the CMR literature. A notable exception is the collection of articles directed at a critical analysis of projects, which led to a number of related publications in 2007 and 2008. We take as an example the 2007 article by Hodgson and Cicmil.

In their article, Hodgson and Cicmil set out their purpose as critiquing the bureaucratic approach to project management associated with formal knowledge tools and, specifically, with the Project Management Body of Knowledge (PMBoK), a compilation of standards sponsored by the Project Management Institute. At the core of their paper is an argument that, by focusing project manager attention on formal, impersonal criteria, knowledge tools such as the PMBoK increase the oppression and exploitation of workers [intellectual insight]. Their hope is to use this critique to facilitate the development of knowledge tools that are flexible, adaptable, reflexive, democratic, and informal. In doing so, they project an alternate vision [purpose]. Theoretically the paper rests on the sociology of standards. In keeping with qualitative approaches, it adopts a constructivist ontology, arguing that standardised formal knowledge produces projects (rather than passively describing them) [phenomenon].

The analysis itself involves a discourse analysis of the PMBoK, thus combining provoking and comprehending purposes. This, however, introduces a disjuncture between the paper's claims and its data analysis. By limiting the analysis to the formal text, the paper fails to provide empirical evidence for the effects on day-to-day management activities, which it claims. This disjuncture illustrates the problem of misplaced modesty – or rather immodesty – signalled in the opening of this chapter. The paper raises important issues of power and power relations that are often ignored in CMR research. But its argument calls for further empirical research focused on the ways in which formal texts construct employee–employee relations and the theorised power effects. Viewed from the perspective of this discussion, it illustrates a different aspect of representationalism in which the conclusions of a theoretical study are presumed to apply equally across contexts.

Discussion and conclusion

Two arguments and a critique

This chapter set out to reflect on the contribution of qualitative research. Our exploration has led us to two arguments and a critique. The first argument is about the importance of theory and theory development for the contribution of qualitative research to both academic and practical knowledge. The second argument is about a heuristic use of theory, which we see as specific to qualitative research and the key to its contributions. Finally, the critique is about the tendency in qualitative CMR to eschew this approach in favour of representationalism at the expense of theory development and of interesting findings.

The argument about the importance of theory rests on a distinction between consultancy and academic research. Whereas consultants offer solutions to client-set problems and often limit themselves to dominant ways of understanding (or at least to their clients' ways), academic research should challenge received views by offering interesting and surprising insights into current problems. At the heart of this endeavour is a break with common-sense understanding and an 'aha' moment. Theory is the main tool that interpretivist researchers have at their disposal to achieve this task. It is the way in which scholars take an empirical problem and translate it into a phenomenon. It is the source of new and different ordering concepts, which offer a different view of the things that we study. Finally, it is the way in which we come up with interesting accounts or contributions.

Turning to the critique, the discussion above offers a number of criticisms of what we see to be poor practice in the use of theory in qualitative CMR. Most of these involve an attempt to apply a positivist logic to interpretivist or qualitative theorising. This approach is captured by the term 'representationalist'. It points to a tendency to reduce theory to a descriptive model. Explanation or theorising in this approach figures as a list of discrete factors, variables, barriers, or other elements, and analysis involves slotting data into pre-set categories or boxes. These elements are presented as influencing a stated outcome, with evidence often based on a single case, often limited to a repetition of expert or senior management understandings, and often accompanied by claims to accuracy. Theorising in this approach is often reduced to claims of validation, where evidence that the theory 'fits' construction-type cases is treated as a contribution, rather than as an artefact of the research design. Missing from this endeavour is a positivist attempt to statistically model the relations between the elements in the model and an interpretivist interest in the mechanisms and processes by which the elements in the ontological model are constituted and interrelated (and the settings by which they vary). Similarly missing is any reflection on scope conditions that would inform the research's relevance criteria for assessing its wider theoretical contribution beyond the phenomenon under investigation. In making these observations it should be clear that our criticism is not of positivist research per se, which, when done well, offers valuable insights into a range of types of problems. Instead, it is of the misguided attempt to apply positivist types of theorising to qualitative research questions, at the expense of both epistemologies.

Our second argument concerns the interpretivist alternative to representationalism, which we refer to as a heuristic use of theory. In this view, theory is a tool to explore theoretical and empirical problems. Far from describing an empirical situation or even a phenomenon, theories offer a lens through which to see them in a new light. While some interpretivist theories work with basic ontological models of the type of things to look for – as in Chi and Javernick-Will's (2011) focus on national political culture and industry structure – the theory is never limited to the model. Instead, it is about the dynamic relations between the elements in the model, the changing forms which those elements assume, and the effect of context on variations in their expression. In this approach, valuable insights depend on research into the forms which the analytic constructs assume in a particular setting, a consideration of whether already theorised mechanisms and processes adequately capture the observed dynamics and an exploration of what more we can learn from this case. Particularly valuable are the places where the theory does not fit, as it is these places that are most likely to lead to surprising findings and theory development. It is this use of analytic concepts and theorised mechanisms to explore, rather than to describe, that distinguishes a heuristic from a descriptive use of theory.

The example of conservatism in the construction sector helps to illustrate both arguments. The lack of innovation and the conservatism of the construction sector is a well-rehearsed trope. In many studies, researchers echo the dominant professional and policy view that this is down to 'attitudes'. This individualistic explanation is convenient as it places the burden for change on individuals and, specifically, on individual employees. In contrast, both Sminia's (2011) and Löwstedt and Räisänen's (2014) papers offer quite different and, dare we say, more interesting accounts. In both cases, their contribution rests on a heuristic use of middle-range theories to explore empirical problems and an attempt to go beyond already theorised relations and contexts.

As discussed above, Sminia (2011) used the neo-institutional account of institutional tension and professional practice as a lens through which to study conservative behaviour, where practices involve the cluster of meanings, tools, and skills supporting routinised activities, and conservatism is expressed in the persistence of pre-tendering activity. In his study, Sminia produced data about the elements in this ontological model. He looked for evidence of institutional tensions and for practices and the way they served to manage those tensions. He used the mechanisms discussed in the theoretical literature to guide his exploration of his data. His paper explains the persistence of pre-tendering by the failure of EU regulations to address the underlying tensions that pre-tendering had helped to address. It also suggests a very different approach to addressing problems associated with regulatory change and reform more generally. Instead of (only) punishing non-compliance, successful reform depends on recognising and resolving core tensions, inherent in construction practice, that the (now illegal) practice previously helped to mitigate. The study also offers a basis to revisit neo-institutional accounts of how actors manage institutional tensions; specifically, it focuses theoretical attention on the repairing or masking of perceived tensions as conditions for change. In doing so, Sminia offers CMR researchers a different, theoretically informed account for the seeming conservatism of construction professionals. This approach presents opportunities for CMR researchers to explore parallel mechanisms in areas of observed conservatism, including health and safety, gender and diversity, environmentalism, and innovation, to name but a few.

In a second alternative to the 'attitudes' explanation, Löwstedt and Räisänen (2014) set out to explain employee resistance to top-down change initiatives. They used social identity theory and, specifically, its analysis of re-enforcement mechanisms to explore their empirical question. This theory focused their attention on individuals and the way in which references to in-groups and out-groups contribute to organisational lock-in. Löwstedt and Räisänen used this ontological model to collect and organise their empirical data. They asked if comparable mechanisms were at play in their evidence, and they looked for ways in which the specificity of the construction setting – with its multiple layers and divisions, affording multiple bases of social identification – might inflect or nuance already theorised processes. The result is a very different account of the conservatism of construction professionals based on their social identification with their occupations (and much weaker identification with their firms). This finding, in turn, is rich in suggestions for future research into employee reception of a range of strategic initiatives and management tools. Like Sminia's paper, Löwstedt and Räisänen's article uses a heuristic approach to theory (in which the analytic constructs making up their ontological model and already theorised mechanisms) that produces surprising results with the potential for both theory development and for a rich empirical research agenda with CMR.

Take-away points

In writing this chapter, the authors had two audiences in mind: 1) PhD students and less experienced researchers, and 2) experienced CMR researchers and the CMR community as a whole. In closing, we would like to underline the main take-away points which we have tried to make for each.

For PhD students and less experienced researchers, the aim of the chapter is to underline the specificity of interpretivist theory and theoretical work and to warn against a reductionist, misplaced positivist approach to theory. Our hope is that the discussion above will support you both in the critical review of papers and in the development of your own work. We suggest that Sandberg and Alvesson's (2021) six criteria of what counts as a theory offers a useful starting point, although it is only one of a number of useful frameworks (for a valuable alternative, see Cornelissen, 2017). We selected it because it placed the issue of contribution at the forefront and demonstrated the relation between a rigorous use of analytic concepts and methods on the one hand, and the formulation and evaluation of research problems on the other (as evidenced in the emphasis on types of theory/purpose, phenomena, and ordering concepts).

Our hope is that having opened a conversation among CMR researchers on Sandberg and Alvesson's typology, you will critically reflect on the ways in which CMR papers engage with the six criteria of theory. By identifying the purpose or type of theorising at play in different papers, you will be in a better position to critique them. This recognition should help to steer you away from the misguided view of knowledge as an encyclopaedia of facts about different cases and of contribution as the application of a well-tested research design to yet another case. A contribution is always a contribution to a practical problem and to an ongoing conversation about how to understand and explain a broader phenomenon.

Building on this, our hope is that this chapter explains and exemplifies the specificity of qualitative forms of theorising. Sandberg and Alvesson's paper identifies five distinct types of theory, four of which are specific to interpretivist research, and the fifth of which, explanation, is almost always accompanied by one of the other four. Our discussion of five empirical papers was designed to illustrate each of these types. It was also designed to demonstrate the way in which the core components of interpretivist theory can be mobilised for purposes of exploration rather than description. We see this as the core of theoretical work and the key to developing interesting and novel insights that challenge received approaches to industry problems.

Turning to our second audience, established researchers, the chapter contributes to what seems to be an ongoing conversation about the relevance of the rigour vs relevance debate for CMR, ongoing reflections on the nature of our field, and the challenge of how to move towards a more cumulative and thus impactful form of knowledge production.

In terms of the first, we start with the assumption that, as an applied academic field, CMR must engage with multiple audiences, including CMR academics,

academics in other disciplines, policy makers, practitioners, and a wide range of societal groups affected by the empirical problems that we study. We also embrace the wide range of processes of knowledge production currently being explored, including different forms of co-production and interventionist research. But we reject the view that this absolves us of engaging seriously with theory and theory development. And we also challenge the view that academic journals are the place to communicate with non-academic audiences. Instead, we argue that theory and theory development are at the heart of our ability to make interesting and original contributions, and that academic outputs – be it PhD theses or academic journal or conference papers – should be governed by the highest standards of rigour and theoretical engagement, so as to be relevant.

Finally, we express our concern at the relative lack of cumulativity in CMR-specific knowledge and we suggest that more systematic and rigorous engagement with theory and theory development can help to address this problem. In reviewing empirical papers for this issue, we were struck by the extent to which CMR scholars cite one another for empirical information, but not for theoretical insights or contributions (Schweber and Leiringer, 2012). The example of the supposed conservatism in the industry, discussed above, illustrates the point. There are numerous empirical studies which explore conservatism. Some focus on its manifestation in the area of health and safety, others on its manifestation in environmentalism. But almost none of them cite theoretical insights from empirical cases outside of their particular focus. A similar point can be made for studies of collaboration or integration or standardisation.

At the core of this neglect is a failure to theorise on phenomena rather than empirical problems and a failure to engage more systematically with the theoretical contribution of CMR colleagues' papers. As should be obvious from the discussion above, one case does not a theory make. Instead, theory development in qualitative research is necessarily a collective activity in which successive studies, by the same or other scholars, are used to explore the way in which a particular mechanism works out in different contexts. And this depends on much more meaningful engagement with and referencing of one another's work.

Nor do we help ourselves with our very short discussions of directions for future research at the end of our academic papers, most of which suffer from either excessive empiricism or excessive abstraction. The discussion of papers above identifies some of the research agendas which each of the papers described supports. In each case, the agenda stems from an identification of the ontological model, mechanisms, and settings explored in the paper and questions about what those theoretical processes would look like in a different setting. They also involve reflection on the specificity of construction sector organisations and how these characteristics may or may not inflect documented mechanism and processes. Our hope is that this chapter will encourage colleagues to be more explicit about the use of theory in their work and about the theoretical research agendas which their work invites, and that we will engage more with one another's theoretical work in charting our next research projects.

References

Aguinis, H., Shapiro, D. L., Antonacopoulou, E. P., & Cummings, T. G. 2014. Scholarly impact: A pluralist conceptualisation. *Academy of Management Learning and Education,* 13, 623–39.

Alvesson, M., & Sandberg, J. 2013. Has management studies lost its way? Ideas for more imaginative and innovative research. *Journal of Management Studies,* 50, 128–52.

Antonacopoulou, E. P. 2009. Knowledge impact and scholarship: Unlearning and practising to co-create actionable knowledge. *Management Learning,* 40, 421–30.

Aram, J. D., & Salipante Jr., P. F. 2003. Brigdging scholarship in management: Epistemological reflections. *British Journal of Management,* 14, 189–205.

Bansal, P., & Corley, K. 2011. The coming of age for qualitative research: Embracing the diversity of qualitative methods. *Academy of Management Journal,* 54, 233–7.

Bartunek, J. M. 2007. Academic–practitioner collaboration need not require joint or relevant research: Toward a relational scholarship of integration. *Academy of Management Journal,* 50, 1323–33.

Birkinshaw, J., Brannen, M. Y., & Tung, R., L 2011. From a distance and generalizable to up close and grounded: Reclaiming a place for qualitative methods in international business research. *Journal of International Business Studies,* 42, 573–81.

Bourdieu, P., Chambordeon, J.-C., & Passeron, J.-C. 1991. *The craft of sociology.* New York, Walter de Gruyter.

Bourdieu, P., & Wacquant, L. J. D. 1992. *An invitation to reflexive sociology.* Cambridge, Polity Press.

Carton, G., & Philippe, M. 2017. Is management research relevant? A systematic analysis of the rigor–relevance debate in top-tier journals (1994–2013). *M@n@gement,* 20, 166–203.

Chi, C. S. F., & Javernick-Will, A. N. 2011. Institutional effects on project arrangement: High-speed rail projects in China and Taiwan. *Construction Management and Economics,* 29, 595–611.

Chia, R., & Holt, R. 2008. The nature of knowledge in business schools. *Academy of Management Learning and Education,* 7, 471–86.

Colquitt, J. A., & Zapata-Phelan, C. P. 2007. Trends in theory building and theory testing: A five-decade study of the *Academy of Management Journal. Academy of Management Journal,* 50, 1281–1303.

Cornelissen, J. P. 2017. Preserving theoretical divergence in management research: Why the explanatory potential of qualitative research should be harnessed rather than suppressed. *Journal of Management Studies,* 54, 368–83.

Courpasson, D. 2013. On the erosion of 'passionate scholarship'. *Organization Studies,* 34, 1243–9.

Daudigeos, T. 2013. In their profession's service: How staff professionals exert influence in their organization. *Management Studies,* 50, 722–49.

Davis, M. S. 1971. That's interesting! Toward a phenomenology of sociology and a sociology of phenomenology. *Philosophy and Social Science,* 1, 309–44.

Fellows, R., & Liu, A. M. M. 2020. Borrowing theories: Contextual and empirical considerations. *Construction Management and Economics,* 38, 581–8.

Gibbons, M., Limoges, C., Nowotny, H., Schwartzman, S., Scott, P., & Trow, M. 1994. *The new production of knowledge: The dynamics of science and research in contemporary societies.* London, Sage.

Green, S. D., Harty, C., Elmualim, A. A., Larsen, G. D., & Kao, C. C. 2008. On the discourse of construction competitiveness. *Building Research & Information,* 36, 426–35.

Green, S. D., & Schweber, L. 2008. Theorizing in the context of professional practice: The case for middle range theories. *Building Research & Information,* 36, 649–54.

Hodgson, D., & Cicmil, S. 2007. The politics of standards in modern management: Making 'the project' a reality. *Journal of Management Studies,* 44, 431–50.

Jarzabkowski, P., Mohrman, S. A., & Scherer, A. G. 2010. Organization studies as applied science: The generation and use of academic knowledge about organizations. Introduction to the special issue. *Organization Studies,* 31, 1189–1207.

Jasanoff, S. (ed.). 2004. *States of knowledge: The co-production of science and social order.* New York, Routledge.

Kilduff, M., & Kelemen, M. 2001. The consolations of organization theory. *British Journal of Management,* 12, S55–9.

Koch, C., & Schultz, C. S. 2019. The production of defects in construction – An agency dissonance. *Construction Management and Economics,* 37, 499–512.

Kondrat, M. E. 1995. Concept, act and interest in professional practice: Implications of an empowerment perspective. *Social Science Review,* 69, 405–28.

Leiringer, R., & Dainty, A. 2017a. Construction management and economics: On the right to disagree, healthy debates and moving forward. *Construction Management and Economics,* 35, 383–4.

Leiringer, R., & Dainty, A. 2017b. Construction management and economics: New directions. *Construction Management and Economics,* 35, 1–3.

Lindkvist, L. 2005. Knowledge communities and knowledge collectivities: A typology of knowledge work in groups. *Journal of Management Studies,* 42, 1189–1210.

Löwstedt, M., & Räisänen, C. 2014. Social identity in construction: Enactments and outcomes. *Construction Management and Economics,* 32, 1093–1105.

Macintosh, R., Beech, N., Antonacopoulou, E. P., & Sims, D. 2012. Practising and knowing management: A dialogic perspective. *Management Learning,* 43, 373–83.

Macintosh, R., Beech, N., Bartunek, J. M., Mason, K., Cooke, B., & Denyer, D. 2017. Impact and management research: Exploring relationships between temporality, dialogue, reflexivity and praxis. *British Journal of Management,* 28, 3–13.

Makadok, R., Burto, R., & Barney, J. 2018. A practical guide for making theory contributions in strategic management. *Strategic Management Journal,* 39, 1530–45.

Merton, R. K. 1967. *On theoretical sociology.* New York, The Free Press.

Minzberg, H. 1979. *The structuring of organizations,* Englewood Cliffs, NJ, Prentice-Hall.

Nicolai, A., & Seidl, D. 2010. That's relevant! Different forms of practical relevance in management science. *Organization Studies,* 31, 1257–85.
Pettigrew, A. 2001. Management research after modernism. *British Journal of Management,* 12, S61–S70.
Pettigrew, A., & Starkey, K. 2016. The legitimacy and impact of business schools – Key issues and a research agenda. *Academy of Management Learning and Education,* 15, 649–64.
Porter, M. E. 1980. *Competitive Strategy,* New York, Free Press.
Sandberg, J., & Alvesson, M. 2021. Meanings of theory: Clarifying theory through typification. *Journal of Management Studies,* 58, 487–516.
Schweber, L. 2014. Putting theory to work: The use of theory in construction research. Paper presented at the *8th CIDB Postgraduate Conference.* University of the Witwatersrand, Johannesburg, South Africa.
Schweber, L., & Leiringer, R. 2012. Beyond the technical: A snapshot of energy and buildings research. *Research and Information,* 40, 481–92.
Sminia, H. 2011. Institutional continuity and the Dutch construction industry fiddle. *Organization Studies,* 32, 1559–85.
Stinchcombe, A. L. 1991. The conditions of fruitfulness of theorizing about mechanisms in social science. *Philosophy of the Social Sciences,* 21, 367–88.
Suddaby, R. 2014. Why theory? *Academy of Management Review,* 39, 407–11.
Tsoukas, H. 1998. The word and the world: A critique of representationalism in management research. *International Journal of Public Administration,* 21, 781–817.
Van de Ven, A. H. 2007. *Engaged scholarship: A guide for organizational and social research.* Oxford, Oxford University Press.
Van de Ven, A. H. 2018. Academic–practitioner engaged scholarship. *Information and Organization* 28, 37–43.
Whetten, D. A. 1989. What constitutes a theoretical contribution? *Academy of Management Review,* 14, 490–5.

5 Understanding construction sector policy through narrative analysis: a critical perspective

Stuart D. Green and Dilek Ulutas Duman

Introduction

The aim of this chapter is to make the case for narrative methods as a means of gaining insights into construction sector policy. Policy can usefully be understood as 'a set of ideas or a plan of what to do in particular situations that has been agreed to officially by a group of people, a business organization, a government, or a political party' (Cambridge English Dictionary, n.d.). There are, of course, other definitions, but this one is useful in emphasising that policy is by no means the preserve of government. The organisations involved in policy making vary hugely across contexts, and in accordance with the demands of different political economies.

Yet the processes through which construction sector policy is produced rarely attracts the critical attention of aspiring researchers. Even more worryingly, the published outputs of policy makers are often taken almost entirely for granted. There seems to be little interest in any critical analysis of construction sector policy in terms of those who are invited to contribute and those who are ignored. Neither does there tend to be any critical focus on whose interests are being served and whose interests are being subjugated. The allocated role of the academic researcher tends to be forever limited to the provision of evidence in support of predetermined policy. However, the really worrying tendency is that this is the role that researchers too often allocate to themselves. A primary example is provided by the research agenda in support of modern methods of construction (MMC).

The focus of our particular interest lies in the way construction sector policy is constituted in narrative form in published reports. Examples of such reports in the UK include Latham (1994), Egan (1998), Wolstenholme (2009), Farmer

(2016), and the *Construction Playbook* (Her Majesty's Government [HMG], 2020). The contention is that these reports are of interest not only in terms of the issues discussed, but also in terms of how they are scripted. Although our primary focus of interest lies on construction policy within the UK, we are careful to position our arguments within the broader global context. Construction sector policy reports globally are invariably shaped and constrained by the meta narratives of the broader political environment within which they are produced. It therefore becomes necessary to denaturalise the constituent arguments rather than accept them at face value. Important insights can be gained not only on the basis of the issues highlighted, but also from the issues which are ignored. Of particular importance is to understand the underlying plot structure and the roles that are assigned to different players.

But perhaps of key importance is the way we theorise the relationship between the words that are written and the way policy is enacted in practice, otherwise construed as performativity (Czarniawska, 2016). The underlying contention is that narratives are the essential media through which policy and practice are enacted. As such, they are deserving of being treated as research objects in their own right. Of crucial importance is the recognition that policy reports have consequences, not least in terms of the self-identities of those involved. Although narrative approaches remain a minority interest among construction management researchers, they are nevertheless increasingly being taken seriously (e.g. Duman et al., 2019; Löwstedt and Räisänen, 2012; Sergeeva and Green, 2019).

Initially, we focus on the nature of construction sector policy and the extent to which it can be meaningfully understood in terms of its constituent narratives. Thereafter, we describe the essential tenets of narrative research prior to illustrating how they might be mobilised for the purposes of understanding MMC. Finally, we offer some reflections on long-term sectoral change and the role that narratives have played in shaping the identities of those involved.

Towards a narrative perspective

Narratives of construction sector policy

In common with many other countries, there seems to be a perennial crisis within the UK construction sector. At the time of writing, there is much hand wringing within policy circles about the alleged 'toxic and dysfunctional race to the bottom' (cf. Lea, 2021). Farmer (2016) adopts a similar tone in the report

Modernise or Die whereby he refers to the sector as a 'sick and dying patient' while portraying himself as a purveyor of modernity. Such tendencies are by no means new (cf. Fernie et al., 2006; Green, 2011; Murray and Langford, 2003), nor are they limited to the UK. However, there is to date limited research which explores the extent to which such policy narratives have in themselves become part of the problem.

It must further be recognised that construction sector policy has important consequences for broader societal goals relating to the built environment. The way that construction is organised affects a range of broader policy areas, not least health, education, commerce, and infrastructure (Foxell and Cooper, 2015). Yet debates about how the construction process might be better organised can rarely be (if ever) separated from the vested interests of those involved. Hence, the advocated ideas are constrained by the need to be persuasive to a range of audiences. Over time, the narratives which are most persuasive to established power groups become formalised as policy. Their legitimacy is dependent not only on their inherent logic, but also on the extent to which they reflect and reinforce the prevailing political discourse of the time. Construction sector policies hence cannot be separated from the political economies within which they are produced.

Policy narratives also have consequences for the material fabric of the built environment and for the working lives of those involved in its production. There will always be those who are disparaging of any notion of 'narrative', preferring instead to focus on facts. In truth, facts rarely speak for themselves; they only become convincing once positioned within a persuasive narrative (Gabriel, 2000). It follows that construction sector policy is socially constructed over time – and continuously contested – through the medium of narrative. Hence, the way that construction sector policy ebbs and flows over time becomes an important focus of empirical study.

Policy agendas are also important in terms of the way they are invariably used to frame research agendas – with direct implications for funding. The danger is that academic researchers find themselves forever trapped by the need to chase funding opportunities shaped by predetermined policy. Research hence becomes subjugated to the prevailing orthodoxy. Yet much is to be gained by deconstructing the way construction policy narratives are organised, and the grand narratives from which they draw.

Narratives can further be seen as providing an essential means of preserving or reproducing stability within organisations, and for promoting or resisting change (Vaara et al., 2016). The adopted perspective is hence especially pertinent to the domain of construction sector policy, where so much emphasis is given to the supposed need for change, and the way it might more effectively be promoted. Others of course may be more interested in how the advocated changes might be better resisted through the articulation of counter-narratives.

Construction sector policy in the global context

It is important to re-emphasise that construction sector policy cannot easily be disaggregated from the broader political economy within which it is embedded. Construction sector policy within modern liberalised economies undoubtedly differs from that which prevails (or perhaps used to prevail) within centralised command economies. Yet there are a host of differences even within these two oversimplified categories. For example, neoliberal economies such as the UK and US differ starkly from the social market economies of Austria, Germany, and the Netherlands. By definition, social market economies favour greater levels of government intervention and higher levels of social welfare. Albert (1991) famously coined the phrase 'Rhineland capitalism' as a euro-centric counter-position to the 'neo-American' model. Construction sector policy in neoliberal countries tends to emphasise the importance of labour market flexibility in support of market competition. The desired mode of innovation in this context is invariably disruptive. In contrast, construction sector policy in social market economies tends to emphasise the importance of training, worker participation, and collective bargaining. And the desired mode of innovation is incremental rather than disruptive. Nordic countries such as Denmark and Sweden are conspicuously much closer to the Rhineland model than to neoliberal economies such as the UK. But almost all neoliberal economies have notably transitioned over the last 40 years away from their own localised Keynesian-inspired versions of the social market economy. Policy narratives can further be seen to be directly implicated in such transitions, not least those which relate to the construction sector. China of course presents a very different case in that the transition to the espoused 'socialist market economy' has been deliberately encouraged through centralised policy. Similar arguments apply to the transitional economies of the former Soviet bloc – albeit with important differences. The contention is that construction sector policy cannot be understood in isolation from the meta narratives which characterise the broader political economy. It follows that construction sector policy is inherently situated within different contexts; hence, its analysis requires a breath of knowledge which extends beyond that routinely taught in construction management degree programmes.

Notwithstanding the above, few would argue that competitive markets are not important for the purposes of driving efficiency. However, few would likewise argue against the need for such markets to be subject to appropriate regulation. The important point is that the dynamics of market competition invariably play out in different ways in different places. Market competition creates incentives for construction firms to improve 'performance'. But the meaning of 'performance' is subject to continuous renegotiation over time, not least because of ever-changing policy demands. Narrative is the medium through which such negotiations take place.

Policy narratives as sectoral-level strategy making

Construction sector policy can further be meaningfully equated with sectoral-level collective strategy making. But as alluded to previously, construction policy narratives do not exist in isolation; they continuously interact with the institutional and political context within which they are produced (cf. Gabriel, 2000; Holstein and Gubrium, 2012). Different actors come together within the context of the pan-industry institutional structures which prevail at the time – with a greater or lesser degree of involvement from government. The question of which actors are invited to participate is again hugely dependent upon the prevailing socio-political context. But such issues are invariably best understood as transitions within which the narratives of construction sector policy are directly implicated. For example, in Nordic countries it is common practice to include trade unions within the formal mechanisms of policy making. But in more liberalised economies such as the UK, trade unions have long since been marginalised. The Wood (1975) report on *The Public Client and the Construction Industries* comprised the death throes of the UK's post-war tripartite collaboration between government, trade unions, and industry. It is hence representative of a political consensus which no longer exists (Ive, 2003). The advocated support for direct employment rapidly became victim to the changing political climate of the late 1970s. Hence, even quasi-historical construction sector policy reports are worthwhile research objects in themselves, not least as a means of understanding the antecedents of current concerns. They also provide windows into the ongoing processes of policy making.

To persevere with the example of the UK, pan-industry bodies such as the Construction Industry Board (CIB) and its modern-day successor, the Construction Leadership Council (CLC), are in themselves the outcome of previous construction policy narratives. The CIB was born from the recommendations of the Latham (1994) report, only to be unceremoniously terminated in the wake of the Egan (1998) report (Adamson and Pollington, 2006). Hence, the industry structures within which policy is formulated are forever representative of temporal solutions to the perceived policy priorities of the past. They are further indicative of the balance of power among competing interest groups at particular points in time. Construction sectors throughout the world invariably comprise a huge diversity of vested interests who are forever engaged in quasi-political action. The determination of which interest groups are given voices and which are suppressed hence becomes of pivotal importance in the shaping of construction sector policy.

The essential tenets of narrative research

From stories to policies

Narrative approaches are increasingly influential within the broader field of organisation and management studies (Vaara et al., 2016). The core contention is that individuals make sense of the world around them through stories (Polkinghorne, 1988). Hence, the stories which people tell about construction become a worthwhile focus of research in themselves. In essence, stories are understood as temporal constructions through which we seek to connect key events and actors (de la Ville and Mounoud, 2015). Understanding such stories, and the implicit meanings embedded within them, can provide meaningful insights into managerial practices, and hence the ongoing shaping of reality (Czarniawska, 2004; Brown and Thompson, 2013). Such arguments can readily be extended to notions of sectoral-level strategy making.

Of further importance is the associated conceptualisation of organisations as comprising a succession of discursive spaces whereby different actors and interest groups compete for attention through the medium of narrative (cf. Fenton and Langley, 2011). Many spoken stories are often highly anecdotal in nature and lacking a fully developed plot structure; as such, they are usefully labelled 'ante-narratives' (Boje, 2001). Stories can further be understood as the means through which individuals continuously test their tentative interpretations of the events in question through interaction with others. Arguments such as these are readily applicable to the context of construction sector policy. Indeed, construction sector policy narratives can be seen to comprise the temporal outcomes of sectoral-level strategy making.

Despite their essential temporality, anecdotal stories can be understood to coalesce into semi-scripted narratives which serve to link the past, present, and future (Deuten and Rip, 2000; Fenton and Langley, 2011). Given that such semi-scripted narratives are jointly negotiated, they can be taken as a proxy for ongoing processes of collective sensemaking (Garud et al., 2011; Brown and

Thompson, 2013). Of particular interest is the way they utilise literary plot structures for the purposes of continuously constructing subject positions for the various parties involved (Brown and Humphreys, 2003). It is on this basis that narratives become inseparable from ongoing processes of identity work.

The insights which can be derived from formally published policy narratives are of course very different from those which can be gained from the analysis of spoken ad hoc stories (cf. Boje, 2001). Both are deserving of research attention, and they clearly require different approaches to data collection and analysis. An interest in the temporal dynamics of organisations would tend to privilege spoken ad hoc ante-narratives. In contrast, an interest in policy would prioritise narratives which are in some sense 'institutionalised' on the basis of being endorsed by formal industry representative bodies. Narratives in the latter sense can usefully provide direct insights into the domain of policy making and the processes through which policies come into being. In this sense they can be understood as bona fide sources of primary data. It is further notable that formally published policy narratives are often cited for decades afterwards by researchers, seemingly with little understanding of the socio-political context from which they were derived. An obvious example is provided by the Egan (1998) report, which continues to be heavily cited internationally – despite its parochial focus on the UK. This becomes more understandable if it is understood as a source of discursive resources which others can use for the purpose of positioning themselves as harbingers of modernisation. More recently, management consultancies such as McKinsey (2017) have cornered the market for the provision of legitimising narratives in the cause of 'disruptive innovation'. Such narratives are seemingly especially valued within corporate boardrooms. However, the research community has much to gain from subjecting such narratives to critical deconstruction. Within the UK context, there is hope yet that counter-narratives of regulation and compliance may become more persuasive in the wake of the Grenfell Tower disaster.

Alternative theoretical perspectives

In advocating the importance of construction sector policy narratives, it must also be recognised that narrative research is a broad church which comprises an array of theoretical and methodological traditions. Vaara et al. (2016) identify three key streams of narrative research which they label as realist, interpretive, and poststructuralist. The realist tradition essentially sees narratives as *representative* of an assumed external reality. Policy makers are thereby accorded the status of experts, and their views are assumed to be representative of 'how things are'. The danger with such approaches is that researchers too easily accept the role of reflecting and reinforcing the currently dominant narrative as advocated by industry leaders.

In contrast, the interpretive tradition sees narratives as sensemaking mechanisms through which individuals, and groups of individuals, ascribe meaning to their lived experiences. Studies in this tradition often refer to the *constitutive* role of narratives in organisational emergence and becoming (Vaara et al., 2016). There also tends to be a much stronger empirical interest in understanding how dominant policy narratives are mobilised in practice. An important sub-theme relates to the performative nature of narratives in terms of the self-identities of those involved (cf. Brown, 2006; Sergeeva and Green, 2019). Duman et al. (2019) notably adopt such an approach in focusing on the collective self-identity of the Turkish international contracting sector. The data source comprises a sectoral-level strategy text titled *Geography of Contractors* as convened and published by the Turkish Contractors Association (TCA; Tayanç, 2011). The publication offers a quasi-historical account which presents the development of Turkish international contracting as an epic story. The stated purpose is to create a common memory for the Turkish international contracting sector. The TCA thereby aims to provide a point of reference for subsequent generations in the hope that they will replicate the acclaimed successes of the past. The analysis is notable for the emphasis given to the changing geo-political landscape within which Turkish contractors operate. The focus lies on the way selected narrative elements are used for the purposes of creating a heroic identity for the key identified actors.

The third approach to narrative research described by Vaara et al. (2016) is labelled 'poststructuralist'. The focus here lies on problematizing dominant 'grand' narratives that might otherwise be accepted at face value. Researchers within this tradition tend to talk about 'hegemonic representations' as a means of emphasising the voices that are seen to be repressed and silenced. The central research question often relates to how particular narratives become dominant over others, and the way this inevitably privileges established power groups. Of particular relevance is the way macro-level policy narratives can be seen to reproduce dominant values and ideologies (cf. Lyotard, 1979). For example, the default acceptance of innovation as a 'good thing' can be linked directly to the advent of the neoliberal enterprise culture (cf. du Gay and Salaman, 1992). Construction sector leaders in the UK frequently tell stories about innovation, which they have seemingly adopted as an essential part of their self-identity (Sergeeva and Green, 2019). It can further be argued that the macro-level narrative of neoliberalism has had a significant impact on the way construction is organised (Erlich, 2021; Green, 2011), not least because it shapes what gets talked about, and what gets ignored – with direct implications for the global research agenda. Of particular note is the way the macro-level narrative of enterprise casts regulation as a barrier to the supposed panacea of disruptive innovation. This prolonged and celebrated 'war on red tape' is cited by some as being the primary cause of the 2017 Grenfell Tower disaster in London (Hodkinson, 2019). Rather than forever directing blame at the supposed 'old-fashioned dinosaurs' of the construction sector, there is therefore a strong argument that the responsibility for the systemic failures associated with Grenfell lies with the UK policy makers who have held sway over the last 40 years. The expression which comes to mind is 'lions led by donkeys', although poststructuralists would of course argue that such terms can only ever be understood as temporal negotiated constructs.

The above described tendencies are by no means limited to the UK – they are readily apparent internationally. This is especially true of relatively liberalised economies such as Australia, New Zealand, and the USA. Nevertheless, there are always localised nuances which need to be taken into account. But what is worrying is the extent to which construction management researchers so often align themselves entirely uncritically with the all-pervasive narrative of disruptive innovation. Higher status is seemingly forever allocated to the champions of change, rather than those upon whom society depends for ensuring the day-to-day material integrity of the built environment.

Analytical concepts

The argument to date is hopefully sufficient to convince even the most cynical of readers that narratives are worth taking seriously. There are two primary modes of analysis which might usefully be mobilised for the purposes of analysing construction policy narratives: thematic analysis and structural analysis (Holstein and Gubrium, 2012). The former would focus on the underlying themes in the selected policy narrative. However, the latter would focus on how the story is constructed in accordance with a specific plot structure (Riessman, 2008). Phrased slightly differently, a thematic analysis would seek to analyse the content of the narrative (the 'told'), and its role in creating a shared social reality for the benefit of the intended audience. In contrast, a structural analysis would seek to infer meaning from the means of 'telling' (Riessman, 2008). The overriding concerns of a structural analysis would hence be the adopted plot structure and the constituent narrative elements (cf. Czarniawska, 2004). Both modes of analysis are inherently interpretive, with the researcher conceptualised as an active participant in the process of deriving meaning. They are further best understood as being complementary rather than standalone alternatives (Barry and Elmes, 1997).

A thematic analysis would typically commence with an initial process of familiarisation. For a written narrative this would invariably involve multiple readings, followed by the coding of selected sections of text in accordance with their content. There would then be a search for recurring patterns, which might be subsequently developed into tentative themes. It is important to emphasise that the analysis process is often highly iterative, and very much dependent upon the interpretation of the researcher. If multiple narratives are being analysed, there would be an emphasis on similarities and differences, not only in terms of themes, but perhaps also on the use of recurrent words (Brown and Humphreys, 2003). Once the tentative themes have been identified, they would be tested back against the data and revised accordingly. The themes would then be defined and named, ideally with a succinct and engaging label. Thematic analysis of this nature is best described as a process of interpretation which oscillates between the original data and an emerging pattern of designated themes (cf. Polkinghorne, 1988). Some sources on narrative analysis notably distinguish between inductive and deductive approaches to thematic analysis (Braun and Clarke, 2006). An inductive approach allegedly allows the themes to emerge from the data without any strong preconceptions – although precisely how they are supposed to 'emerge' is rarely made clear. More realistically, a deductive approach would involve searching the data for resonance with preconceived themes. Many sources advocate the need for an abductive approach whereby the researcher remains open to unanticipated findings, and is forever willing to revise their pre-conceptions accordingly (cf. Polkinghorne, 1988; Reichertz, 2004).

Other sources distinguish between semantic approaches and latent approaches to thematic analysis (Braun and Clarke, 2006). The former involves analysing the text only in terms of its explicit content. Such an approach risks tending towards reductionism, and would perhaps be more meaningfully described as 'content analysis'. Semantic approaches too often see narratives simply as sources of information vis-à-vis an assumed external reality (cf. Vaara et al., 2016). In contrast, latent approaches seek to take account of the implied subtext and to probe underlying assumptions. The constructivist tradition of narrative analysis goes further by focusing on the processes through which narratives are constructed (Riessman, 2008). It follows that any meaningful interpretation of construction policy narratives would be dependent upon a broader knowledge of the contexts within which they are produced (cf. Czarniawska, 2004). For example, a narrowly-construed semantic analysis of the Egan (1998) report might typically focus on its advocacy of 'modern management methods' as a means of improving efficiency. Such an analysis would risk ignoring the political context within which the report was produced, and the discursive processes through which it was negotiated. In contrast, a more broadly based latent approach to narrative analysis would place the Egan report within the prevailing political context of the time. Rather than accept 'modern methods of management' at face value, the advocacy of such methods might (for example) be taken as an expression of faith in the competitive dynamics of enterprise. Hence, the intended message from the newly elected New Labour government was that there was to be no return to the 'old' Labour policies of nationalisation and demand management (Green, 2011).

A structural approach to narrative analysis would go further in seeking to address the reasons why a narrative is structured in a particular way. The focus of interest lies specifically on the way the plot is structured, how events are defined, and the specific roles that are assigned to the various characters (cf. Czarniawska, 2004; Riessman, 2008). Of crucial importance is the recognition that all such elements are situated within a depicted contextual setting (Holstein and Gubrium, 2012). The analysis would typically focus on the literary plot structure used to create the narrative in question. The key narrative elements would hence be identified in terms of the described events, the participating actors, and the specific actions undertaken – all of which take place within a broader socio-temporal context (Polkinghorne, 1988).

The analysis would further seek to assign meaning to the way the various elements are arranged along a temporal dimension. The chosen arrangement may in some cases infer supposed causalities between the specified narrative elements. The plot structure hence becomes the primary organising concept which serves to transform a chronological array of narrative elements into a meaningful and coherent whole (Gabriel, 2000). Possible plot structures would include well-known tropes, such as the epic, the tragic, the comic, and the romantic. The latter two are much less common in published policy narratives, but they may nevertheless be frequently invoked within informal spoken interpretations. The choice of plot structure infuses the narrative elements with different meanings, thereby serving to build a different relationship between the narrator and the targeted audience (Czarniawska, 2004).

Identifying plot structures

The essential starting point that we might take from the above is the way the narrators of construction sector policy reports create subject positions for themselves as 'heroes'. Many such narratives are also notably underpinned by an epic plot structure. They tend to focus on the achievements of a selected group of heroes in driving the sector towards some supposedly universally desired state of utopia. In contrast, other protagonists are typically cast in the role of villains who are intent on holding the industry back on the basis of their commitment to 'old-fashioned' ways of working. Such a plot structure is readily identifiable in the Egan (1998) report, whereby the advocates of techniques such as 'lean construction' are notably framed as heroes. The implied villains are those who adhere to old-fashioned, adversarial practices, and/or those who continue to resist the iconic cause of disruptive innovation. A similar plot structure is evident in the Farmer (2016) report, whereby the advocates of MMC are cast as heroes with the mission of curing a supposedly 'sick and dying patient'. The utopian end goal in both cases is the mythical notion of a modernised industry. The repetition of the same essential narrative over multiple decades seemingly does little to dent its persuasiveness to each successive generation of would-be policy makers. However, the desired end goal often subtly shifts in accordance with the mood of the times. For example, more recent narratives such as the *Construction Playbook* (HMG, 2020) also aspire to the elusive goal of a zero-carbon built environment.

Particular emotive terms are often reserved for the housing sector whereby the adopted narratives frequently seek to satisfy a spiritual longing for a 'home' in addition to the material needs of 'housing' (Glendinning, 2021). Emotionally laden slogans used to accompany housing policy initiatives in the UK include 'decent housing for all', 'homes for the people', and, most famously, 'homes for heroes'. Such rhetorical devices are undoubtedly used to increase the persuasiveness of the message for those who might otherwise be unconvinced. The most recent example is provided by the 'help to buy' slogan, which plays on the desire of first-time buyers to own their first home while at the same time serving to make housing ever less affordable. In contrast, the common counter-narrative tends to argue that affordable housing is dependent upon government intervention in supply. It hence mobilises a tragic plot structure whereby the utopian vision of social housing has been sacrificed in the cause of commodifying housing as a capital asset.

The narrative analysis of policy sources would further typically focus on the events depicted, the actors involved, and the actions undertaken. For example, the counter-narrative cited above invariably flags the key event in precipitating the UK housing crisis as the introduction of the 'right to buy' policy in the early 1980s. It was on this basis that tenants in local authority housing were encouraged to buy their rented properties at subsidised rates, only for such properties ultimately to find their way into the private rental sector. The adopted plot structures do not of course dictate the way the recommendations are received by different audiences, but they nevertheless play an important role in shaping how the debate is framed. The persuasiveness of the adopted plot structure is hence important in influencing which narratives go on to be dominant, and which are progressively marginalised.

It can further be argued that the adopted plot structure in construction sector policy reports is the negotiated outcome of a process of abductive reasoning on the part of those involved. It is easy to envisage a process whereby a series of conjectures are generated and thereafter tested against the intended audience (cf. Barry and Elmes, 1997). The common usage of selected narrative fragments embedded in text boxes (for example) is suggestive of a process of keeping multiple interests on board. Such polyphonic representations are forever temporal, thereby implying that published reports are only indicative of policy making frozen at a point in time (cf. Boje, 2001). Hence, there is little to be gained from focusing on the supposed substantive content of a particular report in isolation from the temporal content within which it was produced.

Narrative analysis further frequently serves to challenge quasi-realist interpretations whereby time is seen to play out in accordance with a linear scale. An alternative perspective would be to view the 'past' and the 'future' as socially constructed categories which are continuously (re)assigned for the purposes of sensemaking (cf. Sandberg and Tsoukas, 2020). From this point of view, it becomes possible for the past, the present, and the future to co-exist within the same organisation, and even within the same space. Furthermore, policy narratives often re-interpret the past for the purposes of shaping the future (Rhodes and Brown, 2005; Vaara et al., 2016). Hence, the past becomes a malleable object which can be mobilised for the purposes of strategic intent.

In contrast to the commonly adopted epic plot structure, the adoption of a tragic plot structure could feasibly serve to engender compassion and sorrow over the failures of a chosen protagonist. However, it is also important to recognise that narratives often combine aspects of different plot structures for the purposes of creating their own bespoke storyline. The meanings attached to different characters or events are therefore by no means fixed. For example, a hero can just as easily be described as a victim in a different part of the same narrative. For those interested in construction sector policy, it is sufficient to recognise that there are significant insights to be gained by focusing on the narrative structures of selected reports. However, despite the ready availability of rich data, there strangely remains little interest in research of this nature.

Modern methods of construction viewed from a narrative perspective

Recurring dreams of modernisation

Perhaps the most topical narrative around which to structure a discussion is that of MMC. At the time of writing, the UK government has a formal 'presumption in favour' of MMC in government procurement policy (HMG, 2020). Yet the precise definition of MMC remains stubbornly illusive. Even the extent to which MMC is 'modern' is, at best, contestable. The Scandinavian tradition of prefabricated housing notably dates back to the Vikings (Price, 2020). More recent variants include industrialised building (Diamant, 1964; Dietz and Cutler, 1971), system building (Finnimore, 1989; Grindrod, 2013), and offsite prefabrication (Gibb, 1999; Pan et al., 2007). The Construction Industry Council (CIC) in Hong Kong notably prefer to talk about modular integrated construction (MiC), introducing yet another acronym into the mix. The popularity of such methods notably ebbs and flows over time, with as many recorded failures as there are successes (Gibb, 2001). Yet the mainstream narrative of MMC remains relentlessly positive, with little apparent interest in learning the lessons of the past. In recent years the accepted narrative has become heavily intertwined with that of digitalisation, such that MMC becomes inseparable from building information modelling (BIM) and other techno-props, such as robotics, exoskeletons, and 3D printing. We are seemingly forever in tantalising reach of technological utopia.

Searching for meaning

The above described dense fog of terminology notably causes positivist researchers much anguish as they continuously strive for precise definitions. There is seemingly a default expectation that the terminology of MMC should be in some way representative of an assumed external reality. In contrast, narrative researchers would typically seek to understand the way the various terms are mobilised for the purposes of articulating some sort of beguiling strategic vision. Of particular interest would be the way the terminologies of MMC evolve in accordance with the shifting policy landscape. For example, within many locations there is a striking resemblance between current narratives in favour of MMC and those which were previously mobilised in support of industrialised methods during the 1960s. In both cases, the desired end goal is a reduced reliance on skilled labour. Narrative researchers would focus on the way the various reports advocating MMC are scripted, not least in terms of the way roles are implicitly allocated to different actors. Such analysis can usefully be carried out through the desktop study of published reports, but much can also be gained through the use of archival sources. Yet scouring through archival sources to access the policy debates of the past is not something which tends to be done by construction management researchers. The consequence is that published research too often lacks any sort of historical perspective, with a corresponding lack of sensitivity for the changing socio-political context within which policy is enacted.

Prevailing tendencies towards technological optimism

Notwithstanding the above, insights can also be gained by accessing the way the various terminologies of MMC are mobilised by individuals in the ad hoc stories they tell in interaction with others. Individuals may of course adopt different terminologies in response to the expectations of different audiences. Hence, there would be little reason to expect such stories to be consistent over time. Indeed, ante-narratives by definition are essentially temporal. They are perhaps most usefully understood as sensemaking mechanisms through which individuals seek to ascribe meaning to their lived experiences. Such stories can also be meaningfully understood as a means of identity work. Ambitious individuals understandably seek to align themselves with the quasi-evangelical narratives offered by the likes of McKinsey (2020) with the aim of portraying themselves as earnest purveyors of modernity. Others of course strive to create alternative identities for themselves as cynics and resisters.

At the time of writing, the majority of research studies into the supposed merits of MMC rarely extend beyond neo-positivist obsessions with the supposed drivers of time, cost, quality, and productivity (cf. Pan et al., 2007). Indeed, much of the research comprises opinion surveys on levels of adoption, and perceived barriers to implementation (cf. Goodier and Gibb, 2007; Rahman, 2014; Pan and Sidwell, 2011). Irrespective of whether such research studies are qualitative or quantitative, they are too often based on the presumption of an external objective reality which is amenable to being captured by means of the advocated method. In reality, the self-appointed industry champions of MMC only tend to be interested in research which reinforces the preferred meta-narrative of technological optimism. Such sources play an important role in bolstering their sense of identities as harbingers of modernity. And researchers are of course under constant pressures to be 'relevant' to the needs of industry. This seemingly remains true irrespective of the extent to which the espoused needs are shaped by the hubristic outputs of the global consultancy sector. McKinsey are especially notable in promoting the cause of modular construction as a means of 'sector transformation' (Bertram et al., 2019). Strangely enough, they make no mention of the rather chequered history of industrialised construction to which Gibb (2001) alludes. Panaceas are seemingly forever more popular than balanced scholarship.

Researchers from a more critical, poststructuralist perspective would tend to focus on the role of dominant narratives such as MMC in preserving existing power structures. Certainly, it is not difficult to construe the accepted mainstream narrative of MMC as a means of marginalising skilled labour. The perennial narrative of the 'skills crisis' is indicative of a long-standing tendency to commoditise the industry's workforce as passive vassals for narrowly defined skills. The only skills which are valued are seemingly those which serve the narrowly defined cause of productivity (Ness and Green, 2012). Regardless of the precise nuances of the adopted approach, there remains a rich variety of insights to be gained from researching MMC as a narrative rather than a supposed technological fix.

Reflections on industry change

Narratives of seduction

Irrespective of the degree of emphasis on MMC, it is notable that modern policy narratives frequently emphasise the importance of adopting a 'mutually beneficial, open and collaborative approach' (e.g., HMG, 2020). The phraseology is reminiscent of the discredited notion of partnering (cf. Bresnen, 2007). Indeed, it is barely 20 years since partnering was lauded as the solution to the construction sector's supposed adversarial culture. The industry champions at the time, almost without exception, unashamedly aligned themselves with the seductive rhetoric of partnering (Green, 1999). Similar arguments can be made in respect of lean construction, which also notably failed to live up to its acclaimed status as a panacea. But the red herrings of the past are now seemingly forgotten as the 'modernisers' of the present fall over themselves in their hubristic enthusiasm for MMC.

The beguiling narratives of partnering and collaborative working are perhaps best ultimately understood as discursive means of offsetting the risks associated with systemic outsourcing (Green, 2011). It is notable that the narrative of partnering in the end lacked plausibility due to the inherent weight of its internal contradictions (cf. Bresnen, 2007). The more recent narrative of integrated project delivery (IPD) is likely to suffer a similar fate (cf. Walker and Rowlinson, 2019). The irony is that each subsequent generation of policy makers falls victim to the same degree of hubris while remaining entirely uninterested in the failed initiatives of the past. Hence, the task of curating the lessons of the past for the purposes of a better-informed future falls increasingly on the shoulders of the research community. Needless to say, research of this nature is unlikely to be valued in policy circles.

Sectoral transitions

A central component of the accepted narrative of construction improvement is that the sector remains largely unchanged in its working practices since the Second World War. It is, of course, complete nonsense. In the context of the preceding discussion, it is important to remind ourselves that there was a time when UK government ministries, such as Health and Education, contained specialist architectural divisions with their own in-house capability in research and development (Duffy and Rabeneck, 2013). Much of this expertise ultimately found its way into the Property Services Agency (PSA) before being spilt up and outsourced in 1993. Hence public-domain knowledge on issues such as building performance was shifted to the private sector. Knowledge that was previously curated in pursuit of the public good was hence thereafter mobilised in the cause of competitive advantage (Foxell, 2019). Many local authorities also previously nurtured and developed their own directly employed construction workforce. In their heyday, direct labour organisations (DLOs) accounted for up to 17 per cent of UK construction sector output. They have since almost entirely disappeared following the widespread imposition of compulsory competitive tendering (Green et al., 2008). Ironically, the DLOs were rendered uncompetitive, in part as a direct consequence of their sustained investment in skills and training, hence the argument that the relentless policy narrative of competitiveness is directly implicated in the endless perpetuation of the 'skills crisis'. Such trends are again by no means unique to the UK. Indeed, they are broadly synonymous with what is understood by transitioning towards a 'liberalised economy'.

The reification of the project

Returning to the theme of global trends, it is important to emphasise that the systemic trend in outsourcing described above can further be equated with ongoing processes of projectification. Indeed, the projectification of construction is now so routine it is invariably taken for granted. The omnipotent narrative of 'project management' has played an important role in normalising the 'project' as the essential unit of production around which construction is organised. This tendency is well-established within the West. It is also increasingly dominant in the former command economies of Eastern Europe and, to a slightly lesser extent, across Central Asia. It is especially evident in the People's Republic of China as it seeks to transition towards the espoused 'socialist market economy'. Such radical processes of restructuring are invariably mediated by top-down policy narratives, with greater or lesser degrees of resistance in the form of counter-narratives. The end result in all cases has been the redrawing of the boundaries between the state and private sector, with significant implications for the self-identities of those involved.

Policy narratives, hence, have direct consequences for the way that construction sectors are organised, and for the lived experiences of the workforce. Modern construction sector policy reports routinely lament the industry's structural characteristics, not least the dysfunctional extent of fragmentation. However, such characteristics are very often a direct consequence of previous policy choices. There is nothing predetermined about the way the construction sector is organised. Of particular relevance is the concept of the 'projectified self' whereby individuals are conditioned to think in terms of projects and how they can best be delivered (Berglund et al., 2020). Such thinking has, over time, become deeply embedded in the self-identities of construction professionals. It has further been continuously reinforced by prevailing policy narratives. The accumulative effect has been a collective abrogation of responsibility for the issues which transcend the boundaries of individual projects. The abiding assumption is that the market will provide.

But nostalgia should be avoided for the way public sector procurement was organised in the past. In the UK, the 'hard-wired' preference for lowest-cost tendering owes much to the Poulson scandal of the early 1970s (Jones, 2012). The PSA was also marked by its own corruption scandal, which came to light in the early 1980s. It follows that construction utopia is likely to be forever unattainable, and that debates about construction are inseparable from the broader narratives of the political economy. It is nevertheless the responsibility of the research community to ensure that the narratives mobilised by construction policy makers are forever subject to critical scrutiny.

Conclusion

The stated aim of this chapter was to make the case for narrative methods as a means of understanding construction sector policy. It has been argued that narratives are the essential media through which policy and practice are enacted. As such, they are deserving of being treated as research objects in their own right. In providing examples of construction policy reports, we have drawn the majority of our examples from the UK. But we have also taken care throughout to emphasise that the selected examples are indicative of global phenomena. We have further suggested that one of the reasons why UK construction policy reports are cited so widely internationally is that they provide resources for self-appointed harbingers of liberalisation everywhere. The UK may not be 'world-beating' in terms of its competence in construction, but it is seemingly in a class of its own when it comes to producing construction sector policy reports. But such reports are best understood as performative narratives rather than in terms of their supposed substantive content. We further contend that insights can be gained by analysing the plot structures around which policy narratives are organised, and the way in which they morph over time.

More controversially, we have argued that the policy narratives of the past are directly implicated in the much criticised 'toxic and dysfunctional race to the bottom'. This interpretation provides a much more meaningful research framing than forever contending that the sector is some sort of a deviant case (cf. Farmer, 2016). But even if the construction sector were indeed a 'sick and dying patient', it would seem sensible to cast a critical eye on previous prescribed courses of treatment as preserved within policy narratives. There is a case to be made that the supposed cure has become worse than the alleged disease. The evidence is there to be seen for those who are motivated to look.

It has further been argued that strategic narratives about industry change do not emerge from nowhere, but are rooted within particular political economies. Important insights can hence be gained not only on the basis of the issues highlighted, but also from the issues which are ignored. Of particular importance is to understand the underlying plot structure and the roles that are assigned to different players.

An example of the insights which can be gained from narrative analysis has been provided in the discussion of MMC. The advocates of MMC invariably cast themselves as heroes of modernisation. In contrast, other parties are cast as villains who perpetuate outdated practices. Such subject positions invariably form part of the adopted literary plot structure. Others, of course, are free to offer their own interpretations on the basis of the context within which they operate. Certainly, we make no claim to having privileged access to any sort of universal truth. However, the policy failures of the past strangely do not dispel an apparent predisposition for hubris among construction sector policy makers. We would argue that an important role of the research community is to guard against this recurring tendency. History tells us that complex policy questions are rarely solved by simplistic technological fixes. Indeed, overconfidence and arrogance on the part of construction policy makers invariably have catastrophic consequences. Ultimately our claim is straightforward: narratives matter.

References

Adamson, D. M., & Pollington, T. (2006). *Change in the construction industry: An account of the UK construction industry reform movement 1993–2003*. Abingdon: Routledge.

Albert, M. (1991). *Capitalisme contre capitalism*. Paris: Éditions du Seuil. English translation, 1993, *Capitalism against capitalism*, London: Whurr.

Barry, D., & Elmes, M. (1997). Strategy retold: Toward a narrative view of strategic discourse. *Academy of Management Review*, 22(2), 429–52.

Berglund, K., Lindgren, M., & Packendorff, J. (2020). The worthy human being as prosuming subject: 'Projectified selves' in emancipatory project studies. *Project Management Journal*, 51(4), 367–77.

Bertram, N., Fuchs, S., Mischke, J., Palter, R., Strube G., & Woetzel, J. (2019). *Modular construction: From projects to products*. McKinsey & Co. https:// www .mckinsey .com/ capabilities/ operations/ our -insights/ modular -construction -from -projects -to -products

Boje, D. (2001). *Narrative methods for organizational & communication research*. London: SAGE.

Braun, V., & Clarke, V. (2006). Using thematic analysis in psychology. *Qualitative Research in Psychology*, 3(2), 77–101.

Bresnen, M. (2007). Deconstructing partnering in project-based organisation: Seven pillars, seven paradoxes and seven deadly sins. *International Journal of Project Management*, 25(4), 365–74.

Brown, A. D. (2006). A narrative approach to collective identities. *Journal of Management Studies*, 43(4), 731–53.

Brown, A. D., & Humphreys, M. (2003). Epic and tragic tales: Making sense of change. *Journal of Applied Behavioral Science*, 39(2), 121–44.

Brown, A. D., & Thompson, E. R. (2013). A narrative approach to strategy-as-practice. *Business History*, 55(7), 1143–67.

Cambridge English Dictionary. (n.d.). *Policy*. https:// dictionary .cambridge .org/ us/ dictionary/english/policy

Czarniawska, B. (2004). *Narratives in social science research*. London: SAGE.

Czarniawska, B. (2016). Performativity of social sciences as seen by an organization scholar. *European Management Journal*, 34(4), 315–18.

de la Ville, V-I., & Mounoud, E. (2015). A narrative approach to strategy as practice: Strategy making from texts and narratives. In D. Golsorkhi, L. Rouleau, D. Seidl, and E. Vaara (eds.), *Cambridge handbook of strategy as practice*, pp. 247–64. Cambridge: Cambridge University Press.

Deuten, J., & Rip, A. (2000). Narrative infrastructure in product creation processes. *Organization*, 7(1), 69–95.

Diamant, R. M. E. (1964). *Industrialised building: 50 international methods*. London: Iliffe.

Dietz, A. G. M., & Cutler, A. S. (eds.). (1971). *Industrialized building systems and housing*. Cambridge, MA: MIT Press.
Duffy, F., & Rabeneck, A. (2013). Professionalism and architects in the 21st century. *Building Research & Information*, 41(1), 115–22.
du Gay, P., & Salaman, G. (1992). The cult(ure) of the customer. *Journal of Management Studies*, 29(5), 615–33.
Duman, D. U., Green, S. D., & Larsen, G. D. (2019). Historical narratives as strategic resources: An analysis of the Turkish international contracting sector. *Construction Management and Economics*, 37(7), 367–83.
Egan, Sir John (1998). *Rethinking construction*. London: Department of the Environment, Transport and Regions (DETR).
Erlich, M. (2021). Misclassification in construction: The original gig economy. *ILR Review*, 74(5), 1202–30.
Farmer, M. (2016). *The Farmer review of UK construction labour model: Modernise or die*. London: Construction Leadership Council.
Fenton, C., & Langley, A. (2011). Strategy as practice and the narrative turn. *Organization Studies*, 32(9)1171–96.
Fernie, S., Leiringer, R., & Thorpe, T. (2006). Change in construction: A critical perspective. *Building Research and Information*, 34(2), 91–103.
Finnimore, B. (1989). *Houses from the factory: System building and the welfare state*. London: Rivers Oram Press.
Foxell, S. (2019). *Professionalism for the built environment*. Abingdon, UK: Routledge.
Foxell, S., & Cooper, I. (2015). Closing the policy gaps. *Building Research and Information*, 43(4), 399–406.
Gabriel, Y. (2000). *Storytelling in organizations: Facts, fictions and fantasies*. Oxford: Oxford University Press.
Garud, R., Dunbar, R. L. M., & Bartel, C. A. (2011). Dealing with unusual experiences: A narrative perspective on organizational learning, *Organization Science*, 22(3), 587–601.
Gibb, A. G. F. (1999). *Off-site fabrication: Prefabrication, pre-assembly and modularisation*. New York: Wiley.
Gibb, A. G. F. (2001). Standardization and pre-assembly – Distinguishing myth from reality using case study research. *Construction Management and Economics*, 19(3), 307–15.
Glendinning, M. (2021). *Mass housing: Modern architecture and state power: A global history*. London: Bloomsbury.
Goodier, C., & Gibb, A. (2007). Future opportunities for offsite in the UK. *Construction Management and Economics*, 25(6), 585–95.
Green, S. D. (1999). Partnering: The propaganda of corporatism? *Journal of Construction Procurement*, 5(2), 177–86.
Green, S. D. (2011). *Making sense of construction improvement*. Oxford: Wiley-Blackwell.
Green, S. D., Harty, C., Elmualim, A. A., Larsen, G. D., & Kao, C. C. (2008). On the discourse of construction competitiveness. *Building Research & Information*, 36(5), 426–35.
Grindrod, J. (2013). *Concretopia: A journey around the rebuilding of postwar Britain*. London: Old Street.
Her Majesty's Government (HMG). (2020). *The construction playbook: Government guidance on sourcing and contracting public works projects and programmes*. London: Cabinet Office.

Hodkinson, S. (2019). *Safe as houses: Private greed, political negligence and housing policy after Grenfell*. Manchester, UK: Manchester University Press.
Holstein, J. A., & Gubrium, J. F. (2012). *Varieties of narrative analysis*. London: SAGE.
Ive, G. (2003). The public client and the construction industries. In M. Murray and D. Langford (eds.), *Construction reports 1944–98*, pp. 105–113. Oxford: Blackwell.
Jones, P. (2012). Re-thinking corruption in post-1950s urban Britain: The Poulson affair, 1972–1976. *Urban History*, 39(3), 510–28.
Latham, Sir Michael (1994). *Constructing the team. Final report of the government/industry review of procurement and contractual arrangements in the UK construction industry*. London: HMSO.
Lea, R. (2021). Toxic and dysfunctional industry in race to the bottom. *The Times*, 3 May. https://www.thetimes.co.uk/article/toxic-and-dysfunctional-industry-in-race-to-the-bottom-kmd8n7ct9
Löwstedt, M. & Räisänen, C. (2012). Playing back-spin balls: Narrating organizational change in construction. *Construction Management and Economics*, 30(9), 795–806.
Lyotard, J.-F. (1979). *The postmodern condition: A report on knowledge*. Minneapolis: University of Minnesota Press.
McKinsey. (2017). *Reinventing construction: A route to higher productivity*. McKinsey Global Institute.
McKinsey. (2020). *The next normal in construction: How disruption is reshaping the world's largest ecosystem*. McKinsey & Co.
Murray, M., & Langford, D. (eds). (2003). *Construction reports 1944–98*. Oxford: Blackwell.
Ness, K., & Green, S. D. (2012). Human resource management in the construction context: Disappearing workers in the UK. In A. Dainty & M. Loosemore (eds.), *Human resource management in construction: Critical perspectives* (2nd ed.), pp. 18–50. Abingdon, UK: Routledge.
Pan, W., Gibb, A. D. F., & Dainty, A. R. J. (2007). Perspectives of UK housebuilders on the use of offsite modern methods of construction. *Construction Management and Economics*, 25(2), 183–94.
Pan, W., & Sidwell, R. (2011). Demystifying the cost barriers to offsite construction in the UK. *Construction Management and Economics*, 29(11), 1081–99.
Polkinghorne, D. E. (1988). *Narrative knowing and human sciences*. Albany: State University of New York Press.
Price, N. (2020). *Children of ash and elm: A history of the Vikings*. London: Allen Lane.
Rahman, M. M. (2014). Barriers of implementing modern methods of construction. *Journal of Management in Engineering*, 30(1), 69–77.
Reichertz, J. (2004). Abduction, deduction and induction in qualitative research. In U. Flick, E. Von Kardoff, & I. Steinke (eds.), *A companion to qualitative research*, pp. 159–164. London: SAGE.
Rhodes, C., & Brown, A. D. (2005). Narrative, organizations and research. *International Journal of Management Reviews*, 7(3), 167–88.
Riessman, C. K. (2008). *Narrative methods for the human sciences*. Thousand Oaks, CA: SAGE.
Sandberg, J., & Tsoukas, H. (2020). Sensemaking reconsidered: Towards a broader understanding through phenomenology. *Organization Theory*, 1, 1–34.
Sergeeva, N., & Green, S. D. (2019). Managerial identity work in action: Performative narratives and anecdotal stories of innovation. *Construction Management and Economics*, 37(10), 604–23.

Tayanç, T. (2011). İnşaatçıların coğrafyası: Türk inşaat sektörünün yurtdışı müteahhitlik hizmetleri serüveni [Geography of contractors: The adventure of the Turkish construction industry in international contracting services]. Ankara: Tarih Vakfı (in Turkish).

Vaara, E., Sonenshein, S., & Boje, D. (2016). Narratives as sources of stability and change in organizations: Approaches and directions for future research. *Academy of Management Annals*, 10(1), 495–560.

Wood, Sir Kenneth. (1975). *The public client and the construction industries*. London: HMSO.

Walker, D. H. T., & Rowlinson, S. (eds). (2019). *Routledge handbook of integrated project delivery*. London: Routledge.

Wolstenholme, A. (2009). *Never waste a good crisis*. London: Constructing Excellence.

6 Beyond the boundary and scale of the construction project

Paul W. Chan

Introduction

For many decades now, there have been longstanding calls to transform construction, often by promoting recipes for change that favour technological solutions borrowed from other sectors, such as manufacturing (e.g. Östergren and Huemer, 1999; Murray and Langford, 2003; Fernie et al., 2006; Green et al., 2008). For example, following the Global Financial Crisis of the late 2000s, automation, off-site fabrication, and digitalisation were among the solutions proposed in the fourth industrial revolution that promised improvements in construction productivity and performance (e.g. World Economic Forum, 2016). Such calls continue to be reiterated during the COVID-19 pandemic; the McKinsey Global Institute's *The Next Normal in Construction*, for instance, suggested that scaling up standardisation, modularisation, and digital technologies would accelerate the transformation of global construction from a fragmented project-based industry to a market-based industrialised process (Ribeirinho et al., 2020). Yet, despite these repeated recommendations for technological advancements as panaceas to improve the industry, the problems of fragmentation and the resultant poor performance of the industry persist.

In this chapter, it is argued that this persistence can be attributed in part to the narrowness with which boundaries are drawn on what matters to construction management researchers. By placing a dogged emphasis on the 'project' and the object of delivering the 'building', prevailing scholarship in construction management research has largely been concerned with understanding and developing ways of producing the built environment on schedule and on budget. In turn, underpinning this focus on productivity is the implicit assumption that growth matters. Thus, in transforming the production of the built environment, recommendations to invest in production technologies and business models to digitalise and industrialise construction are often framed in terms of building more and building quicker, rather than building better.

The aim of this chapter is therefore to seek a broader agenda for construction management researchers that goes beyond the construction project narrowly defined. In this pursuit to widen our research considerations, inspiration is drawn from the (sustainability) transitions literature, which points to a need to consider change in a multilevel context. In so doing, the conceptualisation of projects should move beyond the bounded rationalities of time and cost. Instead, this chapter calls for expanding our spatiotemporal perspectives, which will enable the stretching of our collective imagination to build better. This, it is argued, will also return the field of project studies back to its roots as a source of post-bureaucratic creative exploration and innovation.

Changing agendas in construction management research: from utilitarian instruments to questioning the values of construction

Despite recognition that the management of construction is a sociotechnical process where practices are a result of the interdependencies and interactions between humans and material objects (e.g. Harty, 2005; Schweber and Harty, 2010; Sackey et al., 2015; Pirzadeh et al., 2021), construction management research is still predominantly framed in instrumental and technical terms. Take, for instance, the top ten most-cited papers of all time published in *Construction Management and Economics* (Table 6.1); of these, nearly half are about identifying and prioritising factors that contribute to success or failure, which are often couched in terms of time or cost performance. Such papers are typically based on perception-based surveys, which means that these analyses are usually superficial and do not penetrate the depths and complexities of everyday practices of construction management.

In a recent bibliometric analysis of over 3500 studies published in the construction management research field from 2000 to 2020, Bilge and Yaman (2022) identified that the key matters of concern for researchers included building information modelling, information management, scheduling and cost optimisation, lean construction and agile approach, and megaprojects. They also found strong alignment between construction management research and the Project Management Institute's Body of Knowledge (PMBoK) since the PMBoK in its various editions was identified as one of the most influential document in attracting citations from the authors of the studies reviewed.

Table 6.1 List of most-cited and most-downloaded papers in construction management and economics of all time (as of 28 January 2023)

Most-cited papers	Most-downloaded papers
1. The construction industry as a loosely coupled system: implications for productivity and innovation, by Dubois and Gadde (2002)	1. Construction in developing countries, by Ofori (2007)
2. Critical success factors for PPP/PFI projects in the UK construction industry, by Li, Akintoye, Edwards, and Hardcastle (2005)	2. Stakeholder impact analysis in construction project management, by Olander (2007)
3. Partnering in construction: a critical review of issues, problems and dilemmas, by Bresnen and Marshall (2000)	3. Editorial: stakeholder management in construction, by Atkin and Skitmore (2008)
4. Factors influencing construction time and cost overruns on high-rise projects in Indonesia, by Kaming, Olomolaiye, Holt, and Harris (1997)	4. Incorporating the effect of weather in construction scheduling and management with sine wave curves: application in the United Kingdom, by Ballesteros-Pérez, Smith, Lloyd-Papworth, and Cooke (2018)
5. Analysis of factors influencing project cost estimating practice, by Akintoye (2000)	5. From client to project stakeholders: a stakeholder mapping approach, by Newcombe (2003)
6. Sustainable construction: principles and a framework for attainment, by Hill and Bowen (1997)	6. What is engineering construction and why is it important? Towards a research agenda, by Brookes (2012)
7. Construction safety training using immersive virtual reality, by Sacks, Perlman, and Barak (2013)	7. Critical success factors for PPP/PFI projects in the UK construction industry, by Li, Akintoye, Edwards, and Hardcastle (2005)
8. Significant factors causing delay in the UAE construction industry, by Faridi and El-Sayegh (2006)	8. Influence, stakeholder mapping and visualization, by Walker, Bourne and Shelley (2008)

Most-cited papers	Most-downloaded papers
9. Stakeholder impact analysis in construction project management, by Olander (2007)	9. The leadership practices of construction site managers and their influence on occupational safety: an observational study of transformational and passive/avoidant leadership, by Grill, Nielsen, Grytnes, Pousette and Törner (2019)
10. Sustainable construction aspects of using prefabrication in dense urban environment: a Hong Kong case study, by Jaillon and Poon (2008)	10. Diffusion of digital innovation in construction: a case study of a UK engineering firm, by Shibeika and Harty (2015)

The peculiarities of project-based organisation of construction work have long been acknowledged. Ever since Stinchcombe (1959) differentiated between more stable operations in the manufacturing sector that favoured bureaucratic administration and the more variable operations in the construction sector that lend more efficiently to craft or professional-based forms of administration, others have tried to make sense of how construction work is managed in distinct ways when compared to other sectors. For instance, Eccles (1981) argued that, as a consequence of the intensive use of technology, construction work needs to be coordinated in the project-organisational form to deal with complexity, uncertainty, and the need for flexibility. Similarly, when questioning whether it is ever appropriate to treat construction as an 'industry', Groák (1994) also stressed that it would be more productive to regard construction as 'agglomerations of projects – rather than as a discrete industry or a fixed constellation of firms' (p. 287).

Drawing attention to construction as a project-based endeavour has also attracted criticisms. Although, like Groák (1994), Green (2006) acknowledged that the construction sector could offer a template of project-based organising for other economic sectors to follow, he also criticised project management scholarship for being overly 'managerial and unidimensional ... understood in terms of its substantive content' and less concerned with 'the interplay between discourse, human agency and industry structure' (p. 233). Thus, the research agenda has largely been held stable, influenced by the substance of such instrumental standards as the PMBoK, where accounting for and optimising the success factors of time and cost performance matters and where addressing stakeholders' needs are reduced to models of stakeholder management (see Table 6.1).

Reviewing past inquiries into the affairs of the UK construction industry, Langford and Murray (2003) also remarked on how unchangeable recurring themes of performance improvement were framed. While they can see the value of driving improvements in the sector, they argued that the benefits of these improvements are almost always assumed to accrue only to the main members of the construction team (i.e. the client, designers, and construction firms) rather than to the public and society at large. Consequently, Langford and Murray (2003) also noted an ever-growing obsession with and surveillance through benchmarking specific and narrow targets of time and cost savings in construction projects.

Scholars have thus challenged the techno-rational approach to establishing what counts in understanding and improving performance. By reducing the 'project' to the substantive content of time and cost performance, where success factors are assumed to provide the basis for identifying 'best practices', standardised accounts of project management through instruments such as the PMBoK neglect the power of human agency as well as political, ethical, and moral considerations (Hodgson and Cicmil, 2007). As Harvey and Knox (2012) critiqued, building work

> involves a great deal more than simply the execution of a planned process of material transformation. As well as technical expertise, their appearance also requires a force of social and political will which is able to generate and foster the belief that these technologies have a capacity to transform the spaces through which they will pass. (p. 523)

More recent scholarship has started to go beyond technical and rational concerns to shine the spotlight on a more social, values-based agenda for construction management research. Kuitert et al. (2019), for instance, argued for a wider view where the basic project values of time, money and quality are expanded to account for societal values. Kuitert et al. (2019) found that while values relating to efficiency, effectiveness, and the quality of the process are more established and regarded as more important in construction, far less attention is currently paid to values relating to innovation, sustainability, integrity, and transparency. As demands for more sustainable products and services increase, there is a need 'to secure room in projects to be able to manage specific public product-related values during the process and not to restrict themselves beforehand' (p. 274); this can be accomplished through innovative approaches to procurement and partnerships.

Others have also investigated the multiple and often conflicting values that professionals have to negotiate. For instance, Bos-de Vos et al. (2016) distinguished between exchange value, professional value, and use value, and found that while making a profit (i.e. exchange value) was important, architects often traded this off with other values, such as work pleasure and reputation (i.e. professional value) and aesthetical delight for the client (i.e. use value) to sustain long-term competitiveness. Similarly, Ahuja et al. (2020) analysed the discursive strategies that architects deploy when confronting conflicting requirements. They found that architects framed their design interventions in terms of 'best for the client' or 'best for the project' as they walked the tightrope of being seen as valuable contributors to the client's objectives on the one hand, while on the other safeguarding their reputation in improving the built environment for everyone (i.e. beyond the immediate client). It is worth noting that Ahuja et al. (2020) view the 'project' not in utilitarian terms of delivering to time and budget, but as a place where architects can affirm 'their identity as experts that contribute to public good through the considered design of the built environment' (p. 598).

Notwithstanding these efforts to enlarge the scope of construction management research, prevailing scholarship suffers from what Sherratt et al. (2020) called an 'ideological crisis of imagination' (p. 1092). Taking the case of research on public–private partnerships, they noted how existing scholarship in construction management research often treats problems faced in the field as a problem of project efficiency and effectiveness. For them, the emphasis on finding the success factors behind project performance falls short of challenging more fundamental issues that condition construction work. Consequently, the ideological crisis of lacking imagination means that wider political and moral questions around the functioning of markets, the impacts of (neoliberal) policy choices and issues of public welfare and societal value go unexamined in construction management research. As Sherratt et al. (2020) argued, 'Projects and delivery coalitions are not bounded entities, governed only by interior logic or their own emergent properties. It is clear that the execution of projects takes place in, and contributes to, a socio-economic network of interactions and exchanges which extend far beyond their own boundaries' (p. 1096).

In the next section, a case is made for rethinking the 'project' in construction management research to explore how pushing current boundaries can help us move towards more sustainable change.

Rethinking the 'project' in construction management research: towards a multilevel and multi-scalar understanding of the spatiotemporal dynamics of projects

In the preceding section, it was argued that construction management researchers have to date drawn tight boundaries around key objects of research. These revolved around narrow perspectives of investigating project performance, typically aimed at identifying success factors that can contribute to optimising and improving time schedules and cost budgets. There is also a tendency to focus on the benefits of performance improvements for key members of the project team, including the client, designers, and construction companies. And even in relation to construction companies, there is often an overemphasis on the perspectives of the main contractors. In a recent text on the economics of construction, Gruneberg and Francis (2019) remarked that 'Much literature on the construction industry is written from the point of view of main contractors' (p. 71), which glosses over the role and workings of other players, such as subcontractors and specialist contractors, that are instrumental in hiring labour to deliver projects in practice.

The narrow conceptualisation of who and what matters in construction management research is also reflected in the boundaries of construction management practice. Numerous commentators have critiqued industry practice for being too conservative and risk averse to change. Hall et al. (2020), for instance, referred to the mirroring trap where existing industry structures mean that the key actors in established organisational and knowledge boundaries are less likely to embrace disruptive technologies and digital innovation. Loosemore et al. (2021a) also see the construction industry as 'a "closed-shop" with strong perceptions of the ideal construction worker, longstanding industry incumbents, ... strong and long-established existing relationships between existing players and risky procurement and payment practices' (p. 1919). In exploring how project-based firms respond to government initiatives on low-carbon and sustainable construction, Sergeeva and Winch (2020) recounted this quote from a Research and Development Manager at a contracting firm:

> some of the senior leaders in various businesses are so busy fighting fires, operational issues that [they] actually do not take the time to step back to think about long-term vision, long-term goals. If there are long-term visions, long-term goals, they are very much about profit-levels, profit-margins rather than carbon necessarily and things like that. (p. 384)

This short-term thinking is also reflected in Whyte and Nussbaum's (2020) study of the transition between the end of projects and the operational phase of business-as-usual. One of the interviewees in their study reflected on a story in the Heathrow Terminal 4 project, where the difference of perspectives between the project manager of the construction project and the operations manager of the terminal was brought to the fore: on the one hand, the project manager was only concerned about receiving a clear and specific brief so he could get on with the design and construction of the building, while on the other hand, it was difficult for the client (in this case, the operations manager) to be clear about the requirements due to the uncertainties and the ever-changing context of airport operations.

By focussing mainly on the perspectives and experiences of key members internal to the project (i.e. designers, contractors, and the client), construction management researchers have also arguably reinforced the narrow boundaries in accounting for industry practice. Consequently, despite the policy rhetoric to improve the performance of the industry (and not just economic performance, but also social and environmental aspects), the construction industry still seems to be driven by short-term profit-making motives. For instance, Martek et al. (2019) found that despite the maturity of rating systems for sustainable building and the setting of sustainability targets by policy makers, industry practice still falls short due to a number of barriers; these include a lack of demand for sustainable building by end-users and a lack of awareness as to what can be done by the supply chain beyond rating systems. Indeed, simply measuring something does not guarantee the improvement of that something. Similarly, Loosemore et al. (2021a) also found a disconnect between government policy to promote social procurement goals (e.g. of generating more employment for disadvantaged communities) and the enactment of such policy intent through construction firms in practice. Oftentimes, contractors cannot see the value of and connections between delivering the building in a timely and cost-effective manner and the accomplishment of social responsibility through, for example, employing and training non-traditional workers. Thus, in these examples, the intent and purchasing power (particularly of public-sector clients) did not translate to the implementation of environmental and social sustainability goals, due in part to the focus on rational instrumentalism where questions around project performance and the management of risks take precedence.

There is therefore room for a broader conceptualisation of the 'project' in construction management research. In the following sub-sections, I will explore the potential for expanding this conceptualisation, and the implications for construction management research and practice. In so doing, I will draw on critical readings in the field of project studies and in sustainability transitions to emphasise the wider multilevel and multi-scalar contexts in which projects in construction are accomplished.

Recapturing the roots of project management as a creative and innovative field

Project management has been described by Hodgson (2004) as a 'technology of post-bureaucratic control ... offering management techniques able to cope with the discontinuity, flexibility and fluidity of work roles, the permeability of organizational boundaries, and the constancy of change' (p. 85). Yet, at the same time, the capacity to deal with flexibility and change is set in tension against managerial desire for bureaucratic control where the structure of performance measurement and benchmarks provide a sense of efficiency and predictability. Thus, Hodgson (2004) shines the spotlight on the paradox of project management in its ability to creatively deal with change on the one hand while emphasising the need to do so in a controlled manner on the other.

In finding the lost roots of project management, Lenfle and Loch (2010) revisited the Manhattan Project of the 1940s and the ballistic missiles development projects, Atlas and Polaris. They traced how the success of these programmes could be attributed to exploration and experimentation, and by building redundancies through maintaining concurrent parallel strategies – strategies that would be dismissed as too time consuming and costly by modern project standards. They also noted that, in the 1960s, the formalisation of the project management profession coincided with a shift towards more planning and control. Taking the US Department of Defense as a case, they found that more centralised decision-making coupled with a change towards fixed-price contracts meant that the phased waterfall approach to planning projects became the norm. This also led to the erosion of practices of trial-and-error that are critical in searching for novelty and innovation, especially in exploration projects that are characterised by high levels of uncertainty. According to Lenfle and Loch (2010), the neglect of exploration and experimentation is unfortunate since, in the examples of the Manhattan Project, Atlas, and Polaris, these practices were not only useful in finding new scientific breakthroughs but also ended up saving time in achieving the mission.

Indeed, in Hodgson's (2004) study of delivering strategic change in a telebanking organisation, he also observed the disconnect between actual project practices based on experimenting and learning from failure (which yielded, at times, more effective results) and the milestones-and-procedures-dominated rhetoric of project management control (which served to alienate those on the front line). In recapturing the lost roots of project management, the examples of Hodgson (2004) and Lenfle and Loch (2010) remind us of the innovative potential of projects, which is often eclipsed by the (false) sense of bureaucratic control that has long featured in the body of knowledge that frames the project profession. As Davies (2014), in reviewing the fields of project management and innovation management, stressed, despite clear links between the management of (one-off) projects and innovation in the 1950s, developments of both fields have since followed distinct pathways; even in spite of more recent convergence, these linkages tended to focus on the application of project management to deal with the uncertainties of innovation. As Davies et al. (2018) explained, where innovation management research is characterised by the adaptability model that emphasises the opportunities of uncertainties, project management research is dominated by the optimising model, where the focus lies in creating certainty through the management (or eradication) of negative risks. Thus, as Davies et al. (2018) noted, while both project management and innovation management share a common starting point – that is, the creation of project management procedures to deal with large-scale projects aimed at developing technological breakthroughs in the 1950s – both fields have failed to learn from each other.

There is a need to balance accounts of the 'project' so that projects *as* innovation (rather than projects *for* innovation) can be further examined. In so doing, researchers can open up space to analyse how experimentation, exploration, and establishing redundant practices can contribute to project success. While there have been calls to encourage ambidextrous thinking in organisations (e.g. Eriksson and Szentes, 2017; Nowacki and Monk, 2020), where exploration and exploitation are considered hand in hand, studies in the construction management research field have tended to neglect the role of exploration as the emphasis is placed on exploitation (see Koch-Ørvad et al., 2019, for a recent exception). Thus, by refocussing attention on the practices of exploration in construction projects, research might find ways to harness the innovative potential of projects and improve performance, ways that may be counterintuitive to the rational, optimising paradigm.

Projects as portals of innovation in multilevel (sustainability) transitions

That construction projects can be seen as portals of innovation is not a new idea, even if this has been downplayed by the construction management research field. After all, projects are typically carried out in the context of change. In built environment research, van Bueren and Broekhans (2013) examined how construction projects could be used as instruments in the mainstreaming of niche technologies in green and sustainable building. Drawing on the sustainability transitions literature, they saw green and sustainable building in the broader context of the multilevel perspective. According to the multilevel perspective (Geels and Schot, 2007; Geels, 2011), sustainability transitions require systemic reconfiguration, and breakthroughs in transforming existing sociotechnical systems (e.g. energy, transport, housing) shape and are shaped by the interplay between three nested levels: niche technologies (micro), established regimes (meso), and the contextual landscape (macro). The multilevel perspective was formed in response to deficiencies found in the technological innovation literature; while new technologies, products, and services might go some way to help us become more sustainable, without systemic societal change, technologies alone are limited to driving radical shifts in our consumption and production patterns and practices (see Smith et al., 2010; Loorbach et al., 2017).

Although the multilevel perspective provides the broader context to understand how sociotechnical systems can be transformed to become more sustainable, the emphasis to date has been on examining how niche technologies can generate the disruptive change as new and emerging actor networks alter discourses and agendas, which in turn influences policy and practice (Loorbach et al., 2017). Niches at the micro level can therefore be seen as protected spaces, arenas where time is permitted for exploration and experimentation, for instance, through financial subsidies. Yet, a perennial problem lies in finding pathways through which breakthroughs, both technical and non-technical, can move beyond niche experiments to influence and reshape existing regimes. Regimes entail established structures and dominant practices and are often seen as resistant to change (see Geels, 2014a). To elaborate, Geels (2014b) explained how firms in an industry can often organise resistance by mobilising their customers and suppliers, lobby government and policy makers, or challenge regulations in order to protect the industry regime in terms of an industrial identity, norms, mindset, and technical knowledge and capabilities. Thus, the challenge remains in terms of finding ways to scale up (in numbers), scale out (in scope), and scale deep (in changing behaviours more generally) so that dominant regimes (e.g. of unsustainable practices) can be toppled (e.g. Moore et al., 2015; Loorbach et al., 2020).

Recent scholarship has begun to question the binary framing of the disruptive niche and resistant regime. Cuppen et al. (2019), for instance, studied the development of shale gas in the Netherlands to find that, far from resisting the exploitation of shale gas, it was the regime actors who supported such development. Indeed, while early transitions scholars focussed on emerging players who disrupted at the niche level, there has been growing emphasis on studying the increasing number of experiments that are also done by the more established regime actors (Grin, 2020). This is where projects as a portal for innovation in the context of sustainable transitions have potential for further investigation and for scaling up breakthroughs in practice.

Returning back to van Bueren and Broekhans (2013), for example, they studied the design and construction of a town hall in Leiderdorp in the Netherlands, a green project commissioned and delivered by mainstream actors. Their findings indicate how competing niche (based on ecological and resource efficiency) and regime (based on economic efficiency) principles interacted with one another, and how these transformed in priority over time – with niche principles being more significant at the design competition stage, and the winning tender going to the submission with the lowest-price tender (a common regime response). After signing the contract, it seemed that 'business-as-usual' prevailed. On the technology front, there was a tension between using niche innovative technologies and those that were tried and tested at the regime level. However, van Bueren and Broekhans (2013) also observed this to be a false dichotomy. In reality, the niche technologies were also tried and tested in other niche projects experimented elsewhere, which in turn convinced the architect of the workability of some technologies after conducting study visits abroad. Thus, there is more nuance to the distinction between the experimental niche and the resistant regime, as '[construction] projects can be portals through which innovations are adapted and used and possibly transferred to the regime' (van Bueren and Broekhans, 2013, p. 148).

As mentioned earlier, construction management researchers have focussed mainly on practices of exploitation, reiterating how the industry regime is generally conservative and lacks innovation (e.g. Eriksson and Szentes, 2017; Hall et al., 2020; Liu et al., 2021). Yet, the analysis of van Bueren and Broekhans (2013) raises the possibility of searching for practices of exploration. In a similar fashion, Eriksson and Szentes (2017) also found that the separation between exploration as radical innovation and exploitation as continuous development draws an artificial line; in practice, 'any type of development is related to exploration, no matter how small the fine-tuning is, while the other alternative is to carry on as normal. Hence, the choice is often dichotomised, to develop or not, rather than sophisticated in terms of analytically deciding the extent of development' (p. 499). As will be argued in the next sub-section, there is room for more research that examines and learns from the practices of exploration in and around construction projects.

There have been notable successes of learning from practices of exploration in the literature, especially in the research on megaprojects. Davies et al. (2014), for instance, analysed innovation in the London Crossrail project, a megaproject that was at the time the largest civil engineering project in Europe, and explained how the innovation process also opened up possibilities of learning from previous megaprojects, such as the building of the Channel Tunnel Rail Link, Heathrow Airport Terminal 5, Thames Tideway Tunnel, and the London Olympics Development. Indeed, there have been successive attempts to document lessons learnt, such as the Learning Legacy following the London Olympics Development Programme, which saw industry, government, and academic researchers collaborate to capture 'best practices' (see http://learninglegacy.independent.gov.uk/).[1] Notwithstanding these attempts, the failure to learn from past projects has generally been a persistent problem in construction (see Liu et al., 2021). Furthermore, while these learning legacies have tried to capture 'best practices', these tend to emphasise the guiding principles and *what* was done to implement these principles, rather than the twists and turns of *how* those involved accomplished these in practice. There is therefore room for researchers to provide richer and thicker descriptions of everyday practices to show how explorations are exploited in practice. In so doing, this will also address criticisms that the literature on transitions has often downplayed the power and practices of human agency (see Geels, 2011, 2020; Koistinen and Teerikangas, 2021).

Projectification of society: shifting from a lonely project to a multi-scalar ecosystem

In analysing how sustainable innovations in construction can be facilitated through exploratory projects, Koch-Ørvad et al. (2019) presented a case study of Gamle Mursten, a Danish company that specialises in cleaning and selling reused bricks. As a disruptor in driving the reuse of materials in circular construction, Gamle Mursten faced many obstacles. These included the ability to stimulate demand in the market, gaining access to old bricks, and regulatory hurdles in certifying the structural strength and quality of reused bricks. Koch-Ørvad et al. (2019) found that, to tackle these challenges, Gamle Mursten started six projects concurrently, half of which were aimed at addressing the problem of access to reused bricks (i.e. supply-side problem) and the other half targeted at improving the documentation and certification problem (i.e. to stimulate demand). Thus, in building the ecosystem for reused bricks in Denmark, these parallel projects helped Gamle Mursten secure their position as an intermediary for the supply and demand of reused bricks in the circular economy.

Koch-Ørvad et al.'s (2019) study is noteworthy for a number of reasons. First, when the 'project' is examined in construction management research, there is often a tacit assumption that the project is a 'building' project. Attention is then placed on identifying optimal pathways to ensure that the building is completed on time and on budget. Yet, the analysis of Gamle Mursten shows that alongside the 'building' project, companies can often start other concurrent projects to facilitate change and innovation. This multi-project context is rarely considered in the construction management literature, apart from the research field on megaprojects (though, see Hedborg and Gustavsson, 2020, for a recent exception). Second, their study also revealed the practices of confronting and negotiating the existing regime (e.g. by offering alternative value propositions to demolishers, and finding ways to circumvent prevailing certification schemes). In so doing, they also showed how exploration projects can also lead to exploitation (in this case, in terms of increasing the use of reused bricks). Third, it is also notable that the wider institutional context played an important part. Specifically, funding from various schemes by the Danish government and the European Union allowed the creation of the protective space needed to develop this niche technology. Thus, this shines the spotlight on the value of public-sector involvement in driving private-sector innovation, and provides a countervailing argument to the dominant discourse of private-sector efficiency/effectiveness in public–private partnerships (an argument also made by Sherratt et al., 2020).

By broadening the 'project' beyond the single worksite to consider the interrelationships with other parallel (exploration) projects, we are then able to identify new and emerging actors that go beyond the usual suspects of the designer, client, and construction firm (see also Gerding et al., 2021, for a recent analysis of emerging and established actor networks in circular construction). Thus, how researchers conceptualise and categorise projects and how they choose which 'project' to follow can therefore result in substantive differences on the focus taken in the analysis. By focussing on the 'construction' or 'building' project, construction management researchers create a blind spot on the transformative power of what goes on in the broader institutional context outside the boundaries of the project (e.g. Lieftink et al., 2019; Oti-Sarpong et al., 2022).

Traditionally, the project is conceptualised as a one-off task with a definitive start and end goal.[2] In the continued projectification of society, Jensen et al. (2016) offered an alternative conceptualisation as they characterised projects as the human condition. They distinguished the traditional disciplinary society from the project society by examining the relations between activity, space, time, and relations. In the projectified society, activity is open-ended, emerging, and organised through projections into the future rather than through repetitions of past routines. Thus, the project society requires spaces that offer flexibility, where time is not a fixed permanent construct but a temporary passage through and between projects. Relations are thus not fixed and hierarchical, but rather connected through ever-changing networks.

Jensen et al.'s (2016) broadening of the conceptualisation of the 'project' has four significant implications for construction management researchers. First, as already pointed out in Koch-Ørvad et al.'s (2019) study, there is a need to pay more attention to the non-routine exploration activities that can drive progress in firms operating in the construction industry, rather than the routine activities of designing and constructing buildings. Second, while construction has often been regarded as a localised site-based activity, there is also a need to broaden our spatial awareness beyond the local. For instance, in Harvey and Knox's (2012, 2015) study of road construction in Peru, what is seemingly a local road construction project has significant connections with geopolitics in the region (and particularly with neighbouring Brazil) and the power of global finance (see also Rafferty and Toner, 2019; Styhre, 2019). There is thus scope for construction management researchers to expand the spaces of construction to examine how the global and local collide, and to destabilise our a priori assumptions of which actors matter and in what ways by tracing hitherto under-examined associations between stakeholders and actors (see also the example of Schweber and Harty, 2010).

This enlargement on the spatial front can also be found in scholarship that draws attention to the concept of project ecologies (see Grabher, 2002). For example, Hedborg and Gustavsson (2020) zoomed out of the single, lonely project to explore the interrelationships across multiple projects and their implications in building a neighbourhood. By analysing how the municipality, developers, contractors, and suppliers were connected across multiple projects in Stockholm, Sweden, Hedborg and Gustavsson (2020) were able to examine horizontal interdependencies to show how these different actors across multiple projects were often mindful of each other's neighbouring work. This awareness created structural, relational, and cognitive interdependencies to influence critical decisions made within each respective project. One clear example was the use of a construction logistics centre that ensured a more coordinated approach to supplying different project sites simultaneously to meet the high sustainability requirements set out by the municipality. Often, how the decisions made in one project affect and are affected by decisions made in neighbouring projects is missed by construction management researchers who typically focus on the single 'building' project.

Third, there is also a need to extend our temporal perspectives to consider the longer-term sustainability impacts of construction projects. Here, Whyte and Nussbaum (2020) remind us that construction projects do not end at the handover to operations. In a similar vein, Brookes et al. (2017) used an example of a long-term energy infrastructure megaproject to illustrate the complexities of multiple temporalities at play – from the lifecycle of the delivery project with multiple potential scenarios for the operations and decommissioning of infrastructure, to the lifespan of stakeholders and the special purpose vehicle. These examples point to the need to rethink 'project' time as a fixed, bounded, and linear construct. Indeed, it is important to also take into account interdependencies between the management of the construction project and the management of the building asset. After all, it is well known that the cost of operating the built asset far outweighs the costs of designing and constructing the building in the first place. Yet, in focussing on the optimising of time and the costs of construction, decisions are potentially made at the expense of longer-term economic and environmental costs. In analysing how a contractor was exploring a new business model that placed long-term servicing and predictive performance of energy equipment, Robinson et al. (2016) found that decisions were still constrained by thinking about capital expenditure rather than longer-term operating expenditure. Creating longer-term sustainable outcomes can often require investments in new technologies and processes early on. Thus, without stretching our temporal perspectives beyond the timeframe of the project to consider the lifetime costs of the building asset, breakthroughs towards more sustainable outcomes will be limited.

Fourth, there is also a need to develop and understand new relationships, particularly in terms of emerging disruptive players in the ever-evolving actor networks in construction. Take circular construction, for instance: Gerding et al. (2021) mapped out new actor networks across several exemplary projects in the Netherlands to examine the relationships between established (e.g. building contractors) and emerging (e.g. circularity expert) players. While their network analysis indicated that established players still occupy central positions in information exchange, Gerding et al. (2021) also identified instances where involving circularity experts early on in the construction project can help generate end-of-life scenarios that in turn facilitate the implementation of circular designs. They also stressed that the realisation of new ways of working (in their case, on circularity) requires the combination of knowledge from both established and emerging players.

In facilitating the inclusion of emerging actors within established networks in the construction industry, an increasingly significant, if under-examined, role is the role played by intermediaries. In Vihemäki et al.'s (2020) study of how the uptake of sustainable wood construction in multi-storey buildings can be encouraged, the authors highlighted the role intermediaries played in building networks, articulating expectations, promoting learning and exploring, and creating standards. Similarly, Loosemore et al. (2021b) found that intermediaries can play a crucial function in integrating new players unfamiliar to the established regime (in their case, third-sector organisations) within existing project supply chains to drive social procurement implementation in construction projects.

Intermediaries play a vital role in orchestrating relationships to better connect supply and demand in the procurement of innovative solutions (see Edler and Yeow, 2016). An important contemporary development lies in the growth of platforms as critical intermediaries that link demand- and supply-side actors in the ecosystem. Popular examples from outside construction include digital platforms, such as Uber, Airbnb, and Google. Platform thinking is still at a very nascent stage of development in construction management research (e.g. Chan, 2020; Mosca et al., 2020), although this is likely to gain further traction in the near future as platforms have the power to act as intermediaries to integrate both established and emerging players in driving innovative and sustainable solutions. That said, current emphasis appears to focus mainly on more or less established actors and processes on the supply side, with less attention paid to demand-side end-users (see Chan, 2020; Vihemäki et al., 2020).

Concluding remarks

To conclude, prevailing scholarship in construction management has been critiqued in this chapter for its tendency to pursue a narrow research agenda centred on the construction project, and which often frames problems from the perspectives of established actors (namely, designers, clients, and contractors) emphasising the optimisation of a limited set of (mainly time and cost) performance objectives. In so doing, construction management researchers have missed opportunities to take a broader view to engage with wider societal and sustainability transition challenges. These broader challenges demand that the attention of construction management researchers go beyond the boundaries of the construction project.

In this chapter, a more far-reaching reconceptualisation of the construction 'project' was put forward in three ways: (1) projects as a source of creativity and innovation; (2) projects as a portal for breakthroughs in multilevel sustainability transitions; and (3) projects as part of an expansive ecology and multi-scalar ecosystem to drive change. Through these three ways of rethinking the 'project' in construction management research, several under-examined questions have been raised that can provide fertile lines of future inquiry. First, there is the need to shift the emphasis towards understanding and promoting the roles of exploration, experimentation, and adaptation, and to deepen our knowledge of the relationship with (as opposed to the prevailing focus on) exploitation, control, and optimisation. The construction sector has often been regarded as a low-innovation sector, but stressing this point downplays the potential for construction projects to serve as portals for driving innovation. In striking a balance with exploration and exploitation, construction management researchers can then attend to the possibilities of searching for and embedding creative and innovative breakthroughs in the established regime.

Second, by zooming out of micro-level practices that occur within the confines of the construction project to place more emphasis on how the meso-level regime can change or is changing, construction management researchers can pay attention to the incorporation of new actors currently unfamiliar to the established regime. For example, in transforming construction to become more circular or digitalised, there are emerging players, such as circularity advisors, technology firms, and information, product, and digital platforms, among others, that can potentially add value to existing players and processes. Yet, how the emerging players can become integrated with, strengthen, or even outcompete the incumbents remains an under-examined area of research.

Third, and in line with growing attention to new actors beyond the usual suspects of the client, designer, and contractor, construction management researchers can also expand their spatial and temporal considerations of adjacencies and interdependencies. Here, there is room for more research to investigate how life outside the construction project can also shape better, more innovative outcomes in/for the project. Apart from increasing recognition of the need to take into account the asset management life cycle when deciding on the design and delivery of construction projects, researchers can also consider interrelationships with multiple projects, whether directly or indirectly connected within a neighbourhood or an (urban) area. To do so requires construction management researchers to step outside of our knowledge domain to create space for deeper and more meaningful conversations with fields outside of our own.

Notes

1. The learning legacy covered a broad range of themes, including archaeology, town planning, sustainability, procurement, project and programme management, health and safety, design and engineering innovation, systems and technology, equality and inclusion, and transport. The learning legacy has now been archived since October 2016, and it is unclear to what extent the lessons learnt are still kept alive.

2. In a recent essay, Jacobsson and Söderholm (2022) use the terms time bracketing and scope bracketing to illustrate how *Homo projecticus* sets boundaries on beginnings and endings in a projectified society. They argue that bracketing is what conditions actions and facilitates getting things done in projects.

References

Ahuja, S., Nikolova, N. and Clegg, S. (2020) Professional identity and anxiety in architect–client interactions, *Construction Management and Economics*, 38(7), 589–602.

Akintoye, A. (2000) Analysis of factors influencing project cost estimating practice, *Construction Management and Economics*, 18(1), 77–89.

Atkin, B. and Skitmore, M. (2008) Editorial: stakeholder management in construction, *Construction Management and Economics*, 26(6), 549–52.

Ballesteros-Pérez, P., Smith, S. T., Lloyd-Papworth, J. G. and Cooke, P. (2018) Incorporating the effect of weather in construction scheduling and management with sine wave curves: application in the United Kingdom, *Construction Management and Economics*, 36(12), 666–82.

Bilge, E. Ç. and Yaman, H. (2022) Research trends analysis using text mining in construction management: 2000–2020, *Engineering, Construction and Architectural Management*, 29(8), 3210–33.

Bos-de Vos, M., Wamelink, J. W. F. and Volker, L. (2016) Trade-offs in the value capture of architectural firms: the significance of professional value, *Construction Management and Economics*, 34(1), 21–34.

Bresnen, M. and Marshall, N. (2000) Partnering in construction: a critical review of issues, problems and dilemmas, *Construction Management and Economics*, 18(2), 229–37.

Brookes, N. (2012) What is engineering construction and why is it important? Towards a research agenda, *Construction Management and Economics*, 30(8), 603–07.

Brookes, N., Sage, D., Dainty, A., Locatelli, G. and Whyte, J. (2017) An island of constancy in a sea of change: Rethinking project temporalities with long-term megaprojects, *International Journal of Project Management*, 35(7), 1213–24.

Chan, P. W. (2020) Construction in the platform society: new directions for construction management research, in: L. Scott and C. Neilson (Eds.), *Proceedings of the Thirty-Sixth Annual Conference of the Association of Researchers in Construction Management*, Leeds, UK: ARCOM, pp. 396–405.

Cuppen, E., Pesch, U., Remmerswaal, S. and Taanman, M. (2019) Normative diversity, conflict and transition: shale gas in the Netherlands, *Technological Forecasting and Social Change*, 145, 165–75.

Davies, A. (2014) Innovation and project management, in: M. Dodgson, D. Gann and N. Phillips (Eds.), *Oxford handbook of innovation management*, Oxford: Oxford University Press, pp. 625–47.

Davies, A., MacAulay, S., DeBarro, T. and Thurston, M. (2014) Making innovation happen in a megaproject: London's crossrail suburban railway system, *Project Management Journal*, 45(6), 25–37.

Davies, A., Manning, S. and Söderlund, J. (2018) When neighboring disciplines fail to learn from each other: the case of innovation and project management research, *Research Policy*, 47(5), 965–79.

Dubois, A. and Gadde, L.-E. (2002) The construction industry as a loosely coupled system: implications for productivity and innovation, *Construction Management and Economics*, 20(7), 621–31.

Eccles, R. G. (1981) Bureaucratic versus craft administration: the relationship of market structure to the construction firm, *Administrative Science Quarterly*, 26(3), 449–69.

Edler, J. and Yeow, J. (2016) Connecting demand and supply: the role of intermediation in public procurement of innovation, *Research Policy*, 45(2), 414–26.

Eriksson, P. E. and Szentes, H. (2017) Managing the tensions between exploration and exploitation in large construction projects, *Construction Innovation*, 17(4), 492–510.

Faridi, A. S. and El-Sayegh, S. M. (2006) Significant factors causing delay in the UAE construction industry, *Construction Management and Economics*, 24(11), 1167–76.

Fernie, S., Leiringer, R. and Thorpe, T. (2006) Change in construction: a critical perspective, *Building Research and Information*, 34(2), 91–103.

Geels, F. W. (2011) The multi-level perspective on sustainability transitions: responses to seven criticisms, *Environmental Innovation and Societal Transitions*, 1(1), 24–40.

Geels, F. W. (2014a) Regime resistance against low-carbon transitions: introducing politics and power into the multi-level perspective, *Theory, Culture and Society*, 31(5), 21–40

Geels, F. W. (2014b) Reconceptualising the co-evolution of firms-in-industries and their environments: developing an interdisciplinary triple embeddedness framework, *Research Policy*, 43(2), 261–77.

Geels, F. W. (2020) Micro-foundations of the multi-level perspective on socio-technical transitions: developing a multi-dimensional model of agency through crossovers between social constructivism, evolutionary economics and neo-institutional theory, *Technological Forecasting and Social Change*, 152, 119894.

Geels, F. W. and Schot, J. (2007) Typology of sociotechnical transition pathways, *Research Policy*, 36(3), 399–417.

Gerding, D. P., Wamelink, J. W. F. and Leclercq, E. M. (2021) Implementing circularity in the construction process: a case study examining the reorganization of multi-actor environment and the decision-making process, *Construction Management and Economics*, 39(7), 617–35.

Grabher, G. (2002) The project ecology of advertising: tasks, talents and teams, *Regional Studies*, 36(3), 245–62.

Green, S. D. (2006) The management of projects in the construction industry: context, discourse and self-identity, in: D. Hodgson and S. Cicmil (Eds.), *Making projects critical*, Basingstoke, UK: Palgrave Macmillan, pp. 232–51.

Green, S. D., Harty, C., Elmualim, A. A., Larsen, G. D. and Kao, C. C. (2008) On the discourse of construction competitiveness, *Building Research and Information*, 36(5), 426–35.

Grill, M., Nielsen, K., Grytnes, R., Pousette, A. and Törner, M. (2019) The leadership practices of construction site managers and their influence on occupational safety: an observational study of transformational and passive/avoidant leadership, *Construction Management and Economics*, 37(5), 278–93.

Grin, J. (2020) 'Doing' system innovations from within the heart of the regime, *Journal of Environmental Policy and Planning*, 22(5), 682–94.

Groák, S. (1994) Is construction an industry? Notes towards a greater analytic emphasis on external linkages, *Construction Management and Economics*, 12(4), 287–93.

Gruneberg, S. and Francis, N. (2019) *The economics of construction*, Newcastle upon Tyne, UK: Agenda Publishing.
Hall, D. M., Whyte, J. and Lessing, J. (2020) Mirror-breaking strategies to enable digital manufacturing in Silicon Valley construction firms: a comparative case study, *Construction Management and Economics*, 38(4), 322–39.
Harty, C. (2005) Innovation in construction: a sociology of technology approach, *Building Research & Information*, 33(6), 512–522.
Harty, C. (2008) Implementing innovation in construction: contexts, relative boundedness and actor-network theory, *Construction Management and Economics*, 26(10), 1029–41.
Harvey, P. and Knox, H. (2012) The enchantments of infrastructure, *Mobilities*, 7(4), 521–36.
Harvey, P. and Knox, H. (2015) *Roads: an anthropology of infrastructure and expertise*, Ithaca, NY: Cornell University Press.
Hedborg, S. and Gustavsson, T. K. (2020) Developing a neighbourhood: exploring construction projects from a project ecology perspective, *Construction Management and Economics*, 38(10), 964–76.
Hill, R. C. and Bowen, P. A. (1997) Sustainable construction: principles and a framework for attainment, *Construction Management and Economics*, 15(3), 223–39.
Hodgson, D. (2004) Project work: the legacy of bureaucratic control in the post-bureaucratic organization, *Organization*, 11(1), 81–100.
Hodgson, D. and Cicmil, S. (2007) The politics of standards in modern management: making 'The Project' a reality, *Journal of Management Studies*, 44(3), 431–50.
Jacobsson, M. and Söderholm, A. (2022) An essay on 'Homo Projecticus': Ontological assumptions in the projectified society, *International Journal of Project Management*, 40(4), 315–19.
Jaillon, L. and Poon, C. S. (2008) Sustainable construction aspects of using prefabrication in dense urban environment: a Hong Kong case study, *Construction Management and Economics*, 26(9), 953–66.
Jensen, A., Thuesen, C. and Geraldi, J. (2016) The projectification of everything: projects as a human condition, *Project Management Journal*, 47(3), 21–34.
Kaming, P. F., Olomolaiye, P. O., Holt, G. D. and Harris, F. C. (1997) Factors influencing construction time and cost overruns on high-rise projects in Indonesia, *Construction Management and Economics*, 15(1), 83–94.
Koch-Ørvad, N., Thuesen, C., Koch, C. and Berker, T. (2019) Transforming ecosystems: facilitating sustainable innovations through the lineage of exploratory projects, *Project Management Journal*, 50(5), 602–16.
Koistinen, K. and Teerikangas, S. (2021) The debate if agents matter vs. the system matters in sustainability transitions – a review of the literature, *Sustainability*, 13, 2821.
Kuitert, L., Volker, L. and Hermans, M. H. (2019) Taking on a wider view: public value interests of construction clients in a changing construction industry, *Construction Management and Economics*, 37(5), 257–77.
Langford, D. and Murray, M. (2003) Introduction, in: M. Murray and D. Langford (Eds.), *Construction Reports 1944–98*, Oxford: Blackwell, pp. 1–7.
Li, L., Akintoye, A., Edwards, P. J. and Hardcastle, C. (2005) Critical success factors for PPP/PFI projects in the UK construction industry, *Construction Management and Economics*, 23(5), 459–71.
Lenfle, S. and Loch, C. (2010) Lost roots: how project management came to emphasize control over flexibility and novelty, *California Management Review*, 53(1), 32–55.

Lieftink, B., Smits, A. and Lauche, K. (2019) Dual dynamics: project-based institutional work and subfield differences in the Dutch construction industry, *International Journal of Project Management*, 37(2), 269–82.
Liu, Y., Amini-Abyaneh, A., Hertogh, M., Houwing, E.-J. and Bakker, H. (2021) Collaborate to learn and learn to collaborate: a case of exploitative learning in the inter-organizational project, *Engineering, Construction and Architectural Management*, 28(3), 809–30.
Loorbach, D., Frantzeskaki, N. and Avelino, F. (2017) Sustainability transitions research: transforming science and practice for societal change, *Annual Review of Environment and Resources*, 42, 599–626.
Loorbach, D., Wittmayer, J., Avelino, F., von Wirth, T., and Frantzeskaki, N. (2020) Transformative innovation and translocal diffusion, *Environmental Innovation and Societal Transitions*, 35, 251–60.
Loosemore, M., Alkilani, S. Z. and Murphy, R. (2021b) The institutional drivers of social procurement implementation in Australian construction projects, *International Journal of Project Management*, 39(7), 750–61.
Loosemore, M., Denny-Smith, G., Barraket, J., Keast, R., Chamberlain, D., Muir, K., Powell, A., Higgon, D. and Osborne, J. (2021a) Optimising social procurement policy outcomes through cross-sector collaboration in the Australian construction industry, *Engineering, Construction and Architectural Management*, 28(7), 1908–28.
Martek, I., Hosseini, M. R., Shrestha, A., Edwards, D. J. and Durdyev, S. (2019) Barriers inhibiting the transition to sustainability within the Australian construction industry: an investigation of technical and social interactions, *Journal of Cleaner Production*, 211, 281–92.
Moore, M-L., Riddell, D. and Vocisano, D. (2015) Scaling out, scaling up, scaling deep: strategies of non-profits in advancing systemic social innovation, *Journal of Corporate Citizenship*, 58, 67–85.
Mosca, L., Jones, K., Davies, A., Whyte, J. and Glass, J. (2020) Platform thinking for construction, *Transforming Construction Network Plus, Digest Series*, 2, London: University College London. https:// www .ucl .ac .uk/ bartlett/ construction/ sites/ bartlett/files/digest-platform-thinking-for-construction.pdf
Murray, M. and Langford, D. (Eds.) (2003) *Construction reports 1944–98*, Oxford: Blackwell.
Newcombe, R. (2003) From client to project stakeholders: a stakeholder mapping approach, *Construction Management and Economics*, 21(8), 841–48.
Nowacki, C. and Monk, A. (2020) Ambidexterity in government: the influence of different types of legitimacy on innovation, *Research Policy*, 49, 103840.
Östergren, K. and Huemer, L. (1999) Interpretation and translation of industrial recipes: a study of strategic thinking in the Swedish construction industry, *Strategic Change: Briefings in Entrepreneurial Finance*, 8(8), 445–57.
Ofori, G. (2007) Construction in developing countries, *Construction Management and Economics*, 25(1), 1–6.
Olander, S. (2007) Stakeholder impact analysis in construction project management, *Construction Management and Economics*, 25(3), 277–87.
Oti-Sarpong, K., Pärn, E. A., Burgess, G. and Zaki, M. (2022) Transforming the construction sector: an institutional complexity perspective, *Construction Innovation*. https://doi.org/10.1108/CI-04-2021-0071.
Pirzadeh, P., Lingard, H. and Blismas, N. (2021) Design decisions and interactions: a sociotechnical network perspective, *ASCE Journal of Construction Engineering and Management*, 147(10). https://doi.org/10.1061/(ASCE)CO.1943-7862.0002136

Rafferty, M. and Toner, P. (2019) Thinking like capital markets – financialisation of the Australian construction industry, *Construction Management and Economics*, 37(3), 156–68.

Ribeirinho, M. J., Mischke, J., Strube, G., Sjödin, E., Blanco, J. L., Palter, R., Biörck, J., Rockhill, D. and Andersson, T. (2020, June 4) *The next normal in construction: how disruption is reshaping the world's largest ecosystem*, McKinsey and Company, https://bit.ly/constructionnextnormal

Robinson, W. G., Chan, P. W. and Lau, T. (2016) Sensors and sensibility: examining the role of technological features in servitizing construction towards greater sustainability, *Construction Management and Economics*, 34(1), 4–20.

Sackey, E., Tuuli, M. and Dainty, A. (2015) Sociotechnical systems approach to BIM implementation in a multidisciplinary construction context, *ASCE Journal of Management in Engineering*, 31(5). https://doi.org/10.1061/(ASCE)ME.1943-5479.0000303

Sacks, R., Perlman, A. and Barak, R. (2013) Construction safety training using immersive virtual reality, *Construction Management and Economics*, 31(9), 1005–17.

Schweber, L. and Harty, C. (2010) Actors and objects: a socio-technical networks approach to technology uptake in the construction sector, *Construction Management and Economics*, 28(6), 657–74.

Sergeeva, N. and Winch, G. M. (2020) Narrative interactions: how project-based firms respond to Government narratives of innovation, *International Journal of Project Management*, 38(6), 379–87.

Sherratt, F., Sherratt, S. and Ivory, C. (2020) Challenging complacency in construction management research: the case of PPPs, *Construction Management and Economics*, 38(12), 1086–1100.

Shibeika, A. and Harty, C. (2015) Diffusion of digital innovation in construction: a case study of a UK engineering firm, *Construction Management and Economics*, 33(5–6), 453–66.

Smith, A., Voß, J.-P. and Grin, J. (2010) Innovation studies and sustainability transitions: the allure of the multi-level perspective and its challenges, *Research Policy*, 39(4), 435–48.

Stinchcombe, A. L. (1959) Bureaucratic and craft administration of production: a comparative study, *Administrative Science Quarterly*, 4(2), 168–87.

Styhre, A. (2019) Close entanglements: aligning the construction and finance industries, *Construction Management and Economics*, 37(3), 169–78.

van Bueren, E. and Broekhans, B. (2013) Individual projects as portals for mainstreaming niche innovations, in: R. L. Henn and A. J. Hoffman (Eds.), *Constructing green: The social structures of sustainability*, Cambridge, MA: MIT Press, pp. 145–68.

Vihemäki, H., Toppinen, A. and Toivonen, R. (2020) Intermediaries to accelerate the diffusion of wooden multi-storey construction in Finland, *Environmental Innovation and Societal Transitions*, 36, 433–48.

Walker, D. H. T., Bourne, L. M. and Shelley, A. (2008) Influence, stakeholder mapping and visualization, *Construction Management and Economics*, 26(6), 645–58.

Whyte, J. and Nussbaum, T. (2020) Transition and temporalities: spanning temporal boundaries as projects end and operations begin, *Project Management Journal*, 51(5), 505–21.

World Economic Forum (2016) *Shaping the future of construction: a breakthrough in mindset and technology*, in collaboration with the Boston Consulting Group, Geneva: World Economic Forum.

7 Cognition and action in construction project organising

Eunice Maytorena-Sanchez, Natalya Sergeeva and Graham M. Winch

Introduction

In this chapter we consider the central role of uncertainty for cognition and action in construction project organising – specifically, how project practitioners think about the future. We take a cognitive approach to uncertainty (Winch and Maytorena, 2011) in the context of a broader information processing approach to decision-making in organisations inspired by the Carnegie School (Winch, 2015). In doing so, we identify the failure of this approach to connect cognition through to action as one of its limitations and, hence, the main concern of this chapter. We present the Un-Certain Complex Complicated Hidden (UnCoCoH) model to recognise the transition from individual cognition to collective action – where action is future orientated and proactive – and identify the role of narratives for stabilising uncertainty through this transition. Our concern with the relationship between cognition and action is inspired by Schütz (1967) and recent developments in relational sociology (Mische, 2011). This then provides the foundations for working towards the development of a projectivity perspective in construction project organising and advancing a research agenda for this program of research.

To do this we begin by defining construction as a problem in information (Winch, 2015) and revisit the Carnegie School of thought of organisation theory, which places information at the heart of organising (Gavetti et al., 2007, 2012). A key contribution of this work is the focus on how individuals and organisations make decisions under conditions of uncertainty and ambiguity, conceptual ideas which are still relevant today. We therefore move on to explore expected utility theory as the dominant paradigm of decision-making and identify its non-cognitive approach as a limitation. The Carnegie School's interest in cognitive processes to explain individual and organisational mechanisms of decision-making which influence action provide the foundation for research in Managerial and Organisational Cognition (Eden and Spender, 1998), on which we draw on to develop our cognitive perspective. By taking a *cognitive perspective on project uncertainty* (Winch and Maytorena, 2011) we present the value of the concepts of *resolvable* and *radical* uncertainty to construction project organising and explore conceptualisations of uncertainty. However, cognition is not enough, so we move on to focus on action in the context of project leadership. We then present the UnCoCoH model as an action-orientated model. As cognition and action entwine, project leadership can stabilise perceptions of the future through the effective use of narratives. So, we explore the role of narratives in construction project organising. All this paves the way for discussing some theoretical implications drawing on the New York School of relational sociology and outlining a research agenda for a projectivity perspective on construction project organising.

Construction as a problem in information

Research in project organising, generally, and construction project organising, in particular, has evolved over the last 60 years in four main strands with relatively little overlap between them (Winch et al., 2023). These strands are:

- *Projects-as-coordination* through matrix structures (Morris, 1973) and temporary organising (Bryman et al., 1987) to address the fundamental organisational problem of coordination (Puranam, 2018).
- *Projects-as-systems*, drawing on concepts of complexity, life cycles, and homeostasis (Cleland and King, 1968; Ramasesh and Browning, 2014).
- *Projects-as-contracts*, focusing on the commercial interface between the owner and its suppliers (Barnes, 1983; Winch, 2001) and the importance of collaborative working (Pryke, 2020).
- *Projects-as-planning*, developing the tools and techniques (Morris, 1994) of project organising, including the Barnes triangle (Barnes, 1988), schedule and risk management tools, and cost–benefit analysis (Flyvberg and Bester, 2021).

There have been attempts to bridge these separate streams, such as in the theory of the temporary organisation bridging the coordination through temporary organising and systems life cycle concepts (Lundin and Söderholm, 1995), or the three domains model (Winch, 2014), bridging contracts and coordination through matrix organisation concepts. However, we suggest that the concepts of construction as a problem in information can help to make further progress in bringing these streams of research together. For instance, information is at the heart of decision-making and organising, and information processing is the basis for matrix concepts (Galbraith, 1977), transaction cost economics (Williamson, 1975), and systems thinking more generally.

The Carnegie School made significant contributions to our understanding of management and organisations (Gavetti et al., 2007, 2012; Bromiley et al., 2019; Wilden et al., 2019). In particular, they established the information processing approach to organisations based on the contention that the fundamental problem in organising is coordination between organisational units through information flows between them (Puranam, 2018). Thus, March and Simon (1993) argue that, fundamentally, 'organisations process and channel information' (p. 2), while Galbraith (1977) further argues that the basic proposition is that 'the greater the uncertainty of the task, the greater the amount of information that has to be processed' (p. 36). Effective organisations therefore handle uncertainty by processing information through the most appropriate channels. Hence, the Carnegie School's conceptual ideas related to managing and organising, such as decision-making, behaviours, motivations, and their interest in the interaction between the individual and organisation, have been enormously influential and are still relevant today (Bromiley et al., 2019; Wilden et al., 2019). For example, their interest in cognitive processes to explain individual and organisational mechanisms of decision-making that influence action provided the foundation for research in the field of managerial and organisational cognition (Eden and Spender, 1998, cited in Bromiley et al., 2019).

However, research following this tradition has been broad rather than deep. In their reviews, Gavetti et al. (2007, 2012), Wilden et al. (2019), and Bromiley et al. (2019) identify areas of focus which are relevant for researchers in construction project organising and, therefore, this chapter. Gavetti and colleagues (2007, 2012) call for revisiting the Carnegie School's foundations to focus on the neglected elements of decision-making in organisations by incorporating recent developments in individual cognition. Wilden et al. (2019) suggest extending March and Simon's ideas and focusing on two aspects. First, they advocate focusing on new forms of organisations, such as ecosystems, and recognising the importance of systems integration (Hobday et al., 2005) by looking at the relationship between structures and decision-making. Second, scholars should focus on a multilevel analysis to connect the micro-level (individual) and meso-level (organisations) to understand further the way in which these new structures may influence individuals in their information gathering, processing, decision-making, and action. Bromiley et al. (2019) put forward the idea of looking more closely at the concept of 'uncertainty absorption' for its potential to link individual cognition and communication.

These recent reviews present opportunities for research in construction project organising with a focus on decision-making in organisations, and the cognitive processes which influence decision-making and actions under conditions of uncertainty and complexity. Recent work in construction project management has applied these insights from the Carnegie School to temporary construction project organisations by conceptualising the project organisation as an information processing system (Winch, 2010) and project organising as a 'problem' in information (Winch, 2015). Meanwhile, Winch and Maytorena (2011) focus on the cognitive aspects of uncertainty in information processing within the context of project risk management routines. It is to this perspective that we turn after discussing a dominant paradigm in decision-making.

Dominant decision-making paradigm: expected utility theory

An economic perspective has dominated the subject of decision-making and judgement for the past seven decades: Expected Utility Theory (EUT) has been the dominant paradigm for research into decision-making (Schoemaker, 1982). This theory assumes that rational decision-makers act to maximise utility with complete information. However, research has highlighted a number of limitations of EUT's capacity to explain how individuals make decisions, which has driven a growing interest in understanding the cognitive processes involved in decision-making. These include studies on memory, attention, perception, and information processing (Greenwood, 1999; Oppenheimer and Kelso, 2015).

The purpose of decision theory is the study of individuals' choices between alternatives. The area of decision-making under conditions of uncertainty is at its centre. In the 17th century, Blaise Pascal introduced the notion of expected value with his wager on the existence of God; later, in the 18th century, Daniel Bernoulli laid the foundations of probability theory and decision-making science with his work on expected utility. This work, largely based on games of chance, argued that decision-makers should act to maximise expected utility. This approach helped to explain why different individuals assigned different value to the same choices but reduced the ability to predict decision-making behaviour. EUT (von Neumann and Morgenstern, 1944) attempts to address this limitation. This argues that rational individuals act to maximise utility, and act with complete information, where the expected utility of an option is a function of the probability of that option occurring and the expected benefit of that option should it occur. However, this assumes knowledge about objective probabilities for each outcome, and in most decision-making situations in construction project organising this is not possible. Savage (1954) complemented this theory with his work on Subjective Expected Utility (SEU), introducing subjective aspects to the theory of rational decision-making. He argued that probabilities of outcomes are personal or subjective rather than objective, and made clear that probability estimation rests in the mind of the individual and not the state of the world. Later, research by Kahneman and Tversky (1979) found that when decision-makers were presented with identically logical decision choices, different behaviours resulted if these were described as losses rather than gains. From this they developed Prospect Theory, which has been influential in the area of behavioural economics. Kahneman and Tversky's research also found that the elicitation of subjective probabilities suffered from both cognitive biases (Kahneman et al., 1982; Gilovich et al., 2002) and noise (Kahneman et al., 2021). This experimental research showed that individuals use heuristics in many intuitive judgements, which can lead to distinctive biases in decision-making (Gilovich et al., 2002).

This body of work is at the centre of behavioural decision theory (BDT), but it is limited, as the experimental work focuses on outcomes of individual judgements and decision-making behaviour with respect to objective probabilities, and compares them to normative decision models derived from EUT. It therefore fails to consider the inherent subjectivity of our perceptions of future events. In other words, it is non-cognitive because it does not look at knowledge structures or processes (Walsh, 1995; Eden and Spender, 1998). More recently, Oppenheimer and Kelso (2015) argued that increased attention to cognitive processes involved in decision-making is needed and present in the idea of decision-making as information processing. This allows us to connect the meso-structural (Puranam, 2018) contribution of the Carnegie School to the micro-behavioural level of cognition and decision-making. However, the continued focus on choice between alternatives to develop the information processing models is a limitation. A cognitive perspective that recognises the actuality of decision-making in uncertain situations is needed (Alvarez and Porac, 2020) in which 'choice' emerges in the mind of the managers, not between alternatives in the world.

A cognitive perspective

Our cognitive perspective on managing projects under uncertainty is influenced by research in the area of Managerial and Organisational Cognition (MOC; Eden and Spender, 1998; Lant and Shapira, 2001; Huff et al., 2016; Galavan et al., 2017; Galavan and Sund, 2021). MOC seeks to understand how individuals in organisations make sense of their world, how they model reality, and how this influences behaviour, with the aim of improving organisational performance. Specific to the interest of managing projects under uncertainty is the understanding of this field's development – Eden and Spender (1998) provide a comprehensive introduction. Research in this field has been influenced by two perspectives of organisations: organisations as 'information processing systems' (March and Simon, 1958), and organisations as 'interpretation systems' (Neale et al., 2006). Organisations as 'information processing systems' builds on Simon's bounded rationality concept, as well as the developments made by the Carnegie School (Cyert and March, 1963; March and Simon, 1993) on decision-making in organisations. Organisations as 'interpretation systems' build on the work of Weick (1979) and Gioia and colleagues (1991) on sensemaking, sense-giving, and enactment. The focus is on how meaning is shaped by the context and how subsequent actions in turn shape the context. In essence, MOC is concerned with managers' knowledge acquisition processes and the understanding gained through their interactions with their organisational context.

MOC therefore starts from the premise that individuals have limited information processing capacity and that decisions are made under conditions of uncertainty. It dismisses the idea that managerial decisions can be analysed by rational notions of complete information availability and 'logical choice processes', which are the premises of EUT. Furthermore, it argues that individual decision-makers create 'personal models' of the decision situation, different from normative or deterministic models of decision-making.

MOC extends the BDT notions of decision-making by recognising that decision-makers make decisions with limited, sometimes ambiguous, information in an organisational context with varying levels of uncertainty and conflict. MOC views decision-making as a continual process, one that is reflected upon, learned, and socially constructed, and where cognition and action are intertwined. It is from this perspective that we develop our understanding of managing construction projects under uncertainty. The interest in uncertainty is what distinguishes the MOC field from the traditional managerial decision-making body of work, which treats uncertainty as the quantifiable probability distribution around an identified risk (Aven and Renn, 2009; Chapman and Ward, 2011). Therefore, the distinction between the concepts of risk and uncertainty become important for both managerial theory and practice.

A cognitive perspective on managing projects under uncertainty

The importance of understanding the concept of uncertainty in management research and the business context has been the subject of recent debates. For example, Kay and King (2020) introduce the concepts of 'radical' and 'resolvable' uncertainty by drawing on Knight's (1921) and Keynes's (1937) distinction between risk and uncertainty. 'Resolvable uncertainty' is uncertainty that can be captured by a probability distribution of outcomes, and 'radical uncertainty' is uncertainty that cannot be characterised in terms of probabilities. Kay and King present and discuss the implications of 'radical uncertainty' for decision-making in financial, economic, policy, organisational, and individual situations. Within the management theory field, the implications of 'Knightian' uncertainty in managerial situations have been the subject of recent attention with a call to provide more coherence and focus to the construct of uncertainty (Alvarez and Porac, 2020; Griffin and Grote, 2020; Packard and Clark, 2020; Rindova and Courtney, 2020; Arikan et al., 2020; Lampert et al., 2020; Winch, 2023a).

Drawing on the work of Keynes (1937) and Knight (1921) and the theoretical distinction between risk and uncertainty, several researchers have attempted to develop typologies of risk and uncertainty along a continuum that elaborates the dichotomy further to develop models linking cognition and action (Daniel and Daniel, 2018). We summarise the most relevant for construction project organising in Table 7.1, taking the resolvable–radical dichotomy from Kay and King (2020) to provide a conceptualisation of the uncertainty continuum.

Courtney and colleagues (1997) criticise the tendency of strategic managers to 'view uncertainty in a binary way – to assume that the world is either certain, and therefore open to precise predictions about the future, or uncertain, and therefore completely unpredictable' (p. 68). Drawing on case studies of strategic decision-making in larger firms, they encourage managers to think about the business environments they face in more creative ways, and move beyond the certainty–uncertainty dichotomy. They propose that business environments are characterised by one of either *a clear enough future, alternate future,*

Table 7.1 Conceptualising uncertainty

Authors	Resolvable uncertainty ←		Radical uncertainty →	
Courtney et al. (1997)	*A clear enough future*, where traditional strategic planning tools are effective	*Alternate futures*, where discrete scenarios can be identified and analysed using game theory and real-options-type tools	*A range of futures*, where discrete scenarios cannot be identified and scenario-planning-type tools are most appropriate	*True ambiguity*, where there is no basis to forecast the future
DeMeyer et al. (2002)	*Variability*, where projects are affected by a number of deviations from plan which can cumulatively affect the achievement of project objectives	*Foreseen uncertainty*, where events are foreseen which could have a large impact on the project and require contingency planning	*Unforeseen uncertainty*, or 'unk-unks', where events are either completely unforeseen or considered so unlikely that no contingency planning is done	*Chaos*, where assumptions of the nature of the project are unstable
Snowden and Boone (2007)	*Simple* domain, an ordered system is characterised by clear cause-and-effect relationships with repeating patterns of activity, is predictable and can be determined in advanced	*Complicated* domain, also considered an ordered system, is characterised by cause-and-effect relationships that require an analytical approach or expert diagnosis	*Complex* domain is characterised by flux and unpredictability where there is no right answer; cause-and-effect relationships are emergent	*Chaotic* domain is characterised by high turbulence, and much is unknowable; hence, cause-and-effect relationships cannot be identified
Winch and Maytorena (2011)	*Known-known* (risk), where we can identify a possible future event and we can make quantitative inferences from historical data	*Known-unknown*, where we can identify a possible future event but there is no reliable data from which to make quantitative inferences	*Unknown-Known*, where a possible future event has been identified by someone, but it has not yet been revealed to the decision-maker	*Unknown-Unknown*, where a possible future event has not been identified and the decision-maker is in a state of ignorance

a *range of futures*, or *true ambiguity*. Courtney and colleagues apparently treat each of these as discrete states of the world for a particular business environment (e.g. entry into the Indian market), although the examples they give suggest more fluidity than the formal model suggests.

DeMeyer and colleagues (2002; Loch et al., 2006) explicitly address project risk management issues, drawing on cases of larger-scale projects across a number of sectors. For example, in their exploration of the Circored project, they identify the limitations and impact of adopting standard project risk management approaches in one-off, unique projects (Loch et al., 2006). They suggest shifting from risk management to 'uncertainty-based management' as a function of the types of uncertainties it is 'subject to', characterised by: *variability*, *foreseen uncertainty*, *unforeseen uncertainty*, or *chaos*. The 'uncertainty profile' of the project is then the project team's subjective assessment of the relative importance of each type of uncertainty for their project.

Snowden and colleagues (Kurtz and Snowden, 2003; Snowden and Boone, 2007) developed the *Cynefin* framework as a knowledge management perspective. The framework is a sensemaking model which identifies four uncertainty domains defined by the nature of the 'relationship between cause and effect' in that domain, which can be: *simple*, *complicated*, *complex*, and *chaotic*. Simple and complicated domains assume an ordered world, while complex and chaotic domains assume an un-ordered world. The 'un' in un-ordered does not mean disorderly but is used to express a 'paradox, connoting two things that are different but, in another sense, the same' (Kurtz and Snowden, 2003, p. 465). The framework allows the decision-maker to consider the different situations and their dynamics; the range of perceptions and perspectives; and changes to enable the development of a shared understanding and 'decision-making under uncertainty' (Kurtz and Snowden, 2003, p. 468).

A common feature of these frameworks is that they are realist: uncertainty is conceptualised as states of nature rather than states of mind. That is to say, it is treated as an epistemological problem rather than an ontological problem (Winch, 2023a). In contrast, Winch and Maytorena's (2011) cognitive approach to uncertainty on projects combines Savage's fundamental insight regarding the inherently subjective nature of probability and risk, with Knight's/Keynes's distinction between uncertainty and risk (Figure 7.1) and the existence of radical uncertainty. In addition, it introduces an interesting insight from Stephens (2003) regarding unknown-knowns: somebody knows, but is not telling you. The y-axis of certainty and impossibility comes from Keynes (1921), while the x-axis is developed from Galbraith's (1977) definition of uncertainty to the effect that the greater the uncertainty, the greater the

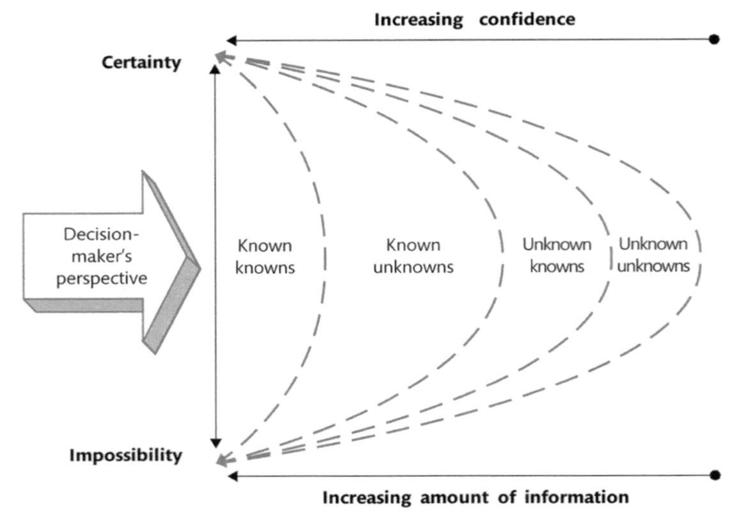

Source: Winch et al. (2022), Figure 2.5.

Figure 7.1 A cognitive model of uncertainty

information processing required to gain confidence (Keynes, 1921) in that information. We use this framing as the foundation for our cognitive perspective on uncertainty. However, one of the limitations is that it does not connect through to action.

From a cognitive perspective to an action perspective

How do these framings shape organisational action? We distinguish action as proactive and, hence, future orientated from behaviour that is reactive and, hence, present- or past-orientated (Schütz, 1967; Winch and Sergeeva, 2022). Much of the literature on the psychological bases of organisational action is derived from pragmatist psychology, particularly the work of William James. Neither of the main traditions in organisation theory identified earlier, which draw on pragmatist psychology, sees the need to provide a categorisation of cognitive states. The information processing approach associated with the Carnegie School stresses the fundamental importance of uncertainty, defined as the lack of information required for a decision, and associated bounded rationality as behaviour under uncertainty. The interpretative approach associated with constructivism (Weick and associates) emphasises equivocality or

Table 7.2 Recommendations for organisational action under conditions of uncertainty

Authors	Managerial and leadership behaviours			
Courtney et al. (1997)	A clear enough future, adapt to the future	Alternate futures, shape the future	A range of futures, reserve the right to play	True ambiguity, shape the future, and reserve the right to play
DeMeyer et al. (2002)	Variability, project managers are trouble-shooters and expeditors	Foreseen uncertainty, project managers are consolidators	Unforeseen uncertainty, project managers are flexible orchestrators and networkers	Chaos, project managers are entrepreneurs
Snowden and Boone (2007)	Simple domain, sense-categorise and respond	Complicated domain, sense-analyse and respond	Complex domain, probe-sense-respond	Chaotic domain, act-sense and respond

ambiguity as cognitive states where there is too much unstructured information rather than uncertainty, but uncertainty is the trigger for sensemaking (Weick, 1995). However, there is a tradition within the organisational behaviour literature of associating differing organisational behaviours with perceptions of threats and opportunities (e.g. Staw et al., 1981). There is much in this literature of relevance to the challenge of managing under uncertainty, but here we will restrict ourselves to reviewing the contributions of those who have attempted to link differing frameworks for understanding risk and uncertainty from Table 7.1 with recommendations for organisational action (Table 7.2).

Courtney and colleagues (1997) identify three strategic postures as responses to different levels of uncertainty:

- '*shape the future*' through playing a leadership role in the industry, which is the preferred strategy when alternate and a range of futures can be identified, but also under true ambiguity, supported by reserving the right to play;
- '*adapt to the future*' through speed, flexibility, and agility, which is the preferred strategy when the future is clear enough but also when adapting to a range of futures; and
- '*reserve the right to play*' by keeping options open is preferred under a range of futures, and under ambiguity.

DeMeyer and colleagues (2002) have developed a typology of managerial styles to respond to differing states of uncertainty:

- under *variability*, project managers are *trouble-shooters and expeditors*, using traditional project management approaches to achieve the project objectives;
- under *foreseen uncertainty*, project managers are *consolidators* who emphasise risk management techniques, sensitivity to environmental changes, and communicating with stakeholders;
- under *unforeseen uncertainty*, project managers are *flexible orchestrators and networkers*, planning iteratively and working with partners through strong, flexible relationships; and
- under *chaos*, project managers are *entrepreneurs*, continually redefining the project while building long-term relationships with stakeholders and others to sense market developments.

Snowden and Boone's (2007) *Cynefin* framework is richest in its recognition for managerial action because it not only proposes appropriate behaviours by the organisational leader, but also identifies some of the dangers of those behaviours if deployed non-reflexively. The leader's job, therefore, is to:

- *sense,* ***categorise****, and respond* when facing simple contexts using best-practice routines, but there is the risk of complacency and failure to spot contexts shifting to another state, which can be mitigated by good communication channels;
- *sense,* ***analyse****, and respond* when facing complicated contexts using evidence-based management, but there is a danger of paralysis by analysis, which can be mitigated by engaging with outside experts;
- ***probe****, sense, and respond* when facing complex contexts using pattern-based management fostering creativity, but there is a danger of impatience and reverting to categorisation and analysis too early; and
- ***act****, sense, and respond* when facing chaotic contexts by providing decisive action to create order while avoiding the danger of the cult of the (successful) leader and ceasing to listen to what is happening.

The cognitive perspective presented in Figure 7.1 lacks an associated action dimension, so we develop one here. All three typologies presented in Table 7.2 place considerable emphasis upon the role of leadership. However, research on leading in project organising has largely addressed who leaders are by focusing on either traits (e.g. Müller and Turner, 2010), or biographies (e.g. Drouin et al., 2021). This approach tends to have the effect of removing leaders from their organisational context. So, we prefer a functional model (Edmondson and Harvey, 2017) that focuses on what leaders do, which we capture in the Project Leadership Model (PLM; Winch et al., 2022) developed from the work of Ancona et al. (2007), which argues that:

- Leadership is pervasive – it is not merely the activity of the Project Director at the most senior level of the project, but takes place at all levels of the project organisation;
- Leadership is personal and developmental, as we learn by doing – we all have our own leadership 'signature' rooted in our capabilities as matured through experience;
- Leadership is incomplete – no one person can excel at all aspects of leadership; each leader has a preferred style and needs a strong team to complement their weaknesses; and
- Within the PLM there are four distinctive and mutually reinforcing leadership processes, which all effective leaders deploy: *projecting, creating, sensemaking*, and *relating*.

By combining the ideas above, we can generate the UNCoCoH model (Figure 7.2). The axes reflect Winch and Maytorena's (2011) cognitive approach (Figure 7.1) and categorizations of uncertainty. These in turn identify a range of decision situations – complicated, complex, hidden, and un-certain – inspired by Snowden and Boone's (2007) domains. We characterise the cognitive state of unknown–unknowns as *Un-Certain*. This formulation is inspired by the middle English word 'undead', meaning 'not quite dead but not fully alive, dead-and-alive' (Oxford English Dictionary). We adopt it here to characterise the paradox (Kurtz and Snowden, 2003) of knowing and not knowing that pervades our knowledge of the future. Like the 'Un-Dead' of Stoker's (1897) *Dracula* who are neither dead nor alive but in a state of unfortunate limbo, the future can be neither true nor false because we can only know the future when it arrives (Aristotle, 1975). But within each situation we can still identify a useful range of tools, warning signs, and predominant leadership processes that can help. The tools and warning signs are derived from the *Cynefin* framework, Ramasesh and Browning (2014), and some that we have found valuable in developing project leaders. These include project pre-mortems (Klein, 2007), causal mapping (Eden and Ackermann, 1998), rich pictures (Checkland, 2001), and stakeholder management (Ackermann and Eden, 2011).

Since the UnCoCoH model helps practitioners focus on action, let's look at the predominant leadership processes within each situation. *Sensemaking* is particularly important for project leaders because of the dynamic and complex (known–unknown) nature of projects. Sensemaking is about how we under-

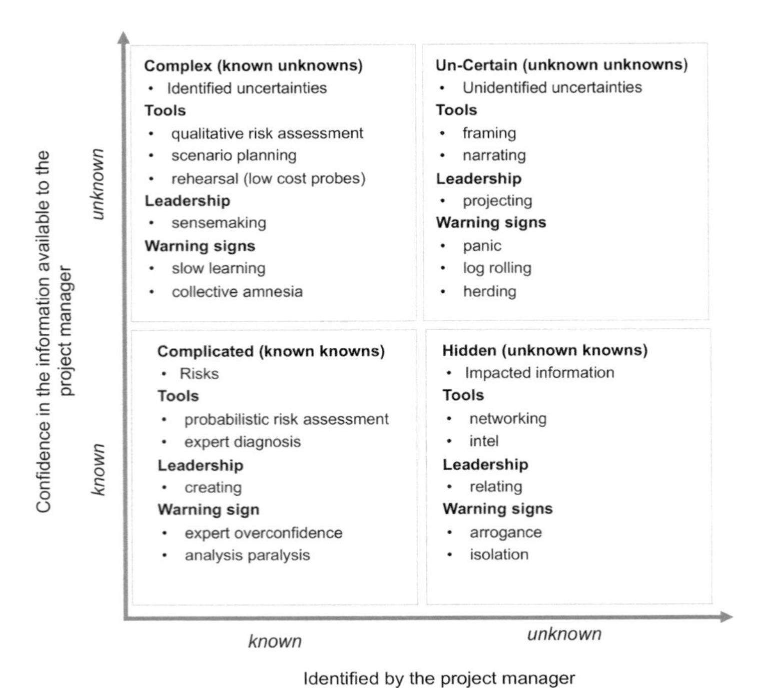

Figure 7.2 UnCoCoH model

stand and interpret the world around us. It is the ongoing, retrospective, social process by which individuals give meaning to their collective experiences (Weick, 1995; Weick et al., 2005). Words and language are a central component of sensemaking and therefore are highly relevant for the other leadership processes. Uncertainty triggers sensemaking, thereby discerning patterns from ambiguity to enable action. It is an active process throughout a project's life cycle. In the context of 'radical uncertainty', asking 'what is going on here?' is a key question for facilitating the exploration of the situation (Kay and King, 2020).

Relating is about building and maintaining trusting relationships within and outside the project organisation, and it is one of the most important activities of the project leader, critical for project shaping, delivery, and in hidden (unknown–known) situations. Relating is both formal, through structured communication and formal processes, and informal, through social networking. According to Ancona et al. (2007), there are three ways to do this: inquiring, advocating, and connecting. The relating dimension of sensemaking is 'sense-giving', directed at external parties whose perceptions are held to be important and, hence, worth influencing (Weick et al., 2005). Therefore, it is a process by which individuals attempt to shape the sensemaking processes of others (Gioia and Chittipeddi, 1991; Maitlis, 2007). For example, project leaders make sense of the environment (sensemaking) and then communicate to others to gain support (sense-giving). Sensemaking is connected to sense-giving in that sensemakers are shaped by 'saying', oriented towards a specific audience, and sensemaking is incomplete without sense-giving, as target audiences may affect sensemakers. Both sensemaking and sense-giving are important processes of relating under conditions of uncertainty.

While *sensemaking* and *relating* are the enabling processes of the PLM, *projecting* and *creating* are the action processes of the PLM. An important part of leading is the process of projecting (Defoe, 1697) – imagining how a project will be developed, progressed, and delivered, and how private and public benefit will be realized through future value. It is at the earlier phases of a project life cycle when project leaders with others are *projecting* the desired future of a project. Project leaders tend to think strategically, and have a vision for a completion of a project. *Projecting* involves creating compelling images of the future; it produces a map of what could be done, what a leader wants the future to be, and it is an ongoing process. Project leaders skilled in *projecting* use stories and narratives to project the desired future. This becomes even more important when in an Un-Certain (unknown–unknown) situation. *Projecting* therefore is about narrating a future (Sergeeva and Winch, 2021). Narratives can be understood as a discursive construction that project leaders use to shape their own individual (sensemaking) and others' understanding (sense-giving), and an outcome of the collective construction of meaning (Brown et al., 2008).

Project leaders use *sensemaking* and *relating* as the enabling processes for *projecting* the project mission and then *creating* how that mission will be delivered. The project narrative therefore ties together the *projecting* and *creating* processes of the PLM (Winch et al., 2022). Creating in the PLM has two dimensions: designing how the project organisation will deliver the outputs, and innovating, which is increasingly recognized as an important activity for project leaders and their teams. Hence, a predominant leadership process is a complicated (known–known) situation. Designing is the process of crafting a temporary project organisation (project team, project structure, project DNA, project identity) that is then communicated internally with the team with the aim of creating a common vision for delivering the owner's mission. Innovating is about problem solving, whether by setting out to advance technology or by combining existing technologies in a novel way to deliver the owner's project mission. Innovating is a step change in best practice that could be a product, process, and service new to the specific context, not necessarily to the world, which could have economic, environmental, or societal benefits for the owner and its stakeholders. Innovating is usually achieved collaboratively across organisations by the people within them, and orchestrating such collaboration is one of the great challenges and opportunities of construction project organising. Collaborating between various individuals, teams, and organisations (owners, project-based firms in the supply chain, and advisors) is the way to innovating (Winch et al., 2022).

The PLM model focuses on the four processes of what project leaders do, but who they are and how they think affects how and what they do as situation-action leaders. We therefore place *judging* at the heart of the PLM and integrate it across all situations in the UnCoCoH model. *Judging*, in broad terms, is about framing, psychology, and experience (Winch et al., 2022). The first is the decision-making frames that shape judgment – models of what good decision-making looks like, on which there is a large literature (e.g. Kahneman et al., 2021); it can be summarized as Dr Optimizer and Dr Skeptic (Klein and Meckling, 1958). The second is the psychological traits capturing both the leader's psychological profile in terms of the Big 5 (Judge et al., 2002), and the leader's ability to empathize with others – emotional intelligence (Mayer et al., 2008). The third captures the leader's experience and ability to draw on that experience to make the most appropriate decisions (Klein, 2017). Project leaders are judging on many aspects of the project throughout the project life cycle based on their sensemaking with others. Hence, *judging* is closely connected with *sensemaking* and *relating*. Project leaders are *judging* what to do and how to do it, and are cognizant that their decisions and actions have important implications for future decisions and actions. Hence, the project leader has a learner mindset. Project leaders' judging influences the project's DNA and the image of the project. *Judging* is hence also connected with the *projecting* and *creating* dimensions of the PLM.

The UnCoCoH model inevitably is a highly synthetic model, and some points should be borne in mind:

- Particular project teams may have differing cognitive states around different situations of the project – the model can either be used in these situations individually, or it may be used to characterise the predominant 'state-of-mind' on the project or decision situation.
- Projects that are perceived to be in an Un-Certain state are unlikely to go ahead – more typically they descend into such a state as unk-unks turn into realities, as on the Cicored project (Loch et al. 2006).

- The leadership styles specified in each category indicate emphasis, rather than suggesting that, for instance, relating is not required for complex project situations.
- There is merit in the project team considering what their project might look like from each of the situations identified ('what if we were a complex project situation?') along the lines suggested by Winter and Szczepanek (2009).
- The cognitive condition of unknown–unknown is not co-extensive with the condition of *unknowable*, although it includes it. Our framework is cognitive; therefore, it takes the perspective of the project manager as decision-maker. It merely describes the condition where the project manager has no conception of the possibility of an event occurring. However, it is also possible that the decision-maker could have known – the condition of 'predictable surprises' (Bazerman and Watkins, 2004).
- It is unclear how this framework relates to Taleb's (2008) analysis of 'black swans'. To take the metaphor literally, the existence of black swans was unknowable to people in Europe until they had discovered Australia, so it is an unknown–unknown in our framework. However, Taleb (2008) appears to use the concept as representing the extreme end of the probability spectrum in terms of occurrence with unpredictable impact, a known–unknown in our framework. We prefer to describe black swans as unk-unks with attitude!
- There is often more data around than is used, and so with a little effort, possible events can be moved from the known–unknown to the known–known category – for instance, probabilistic data on O-ring failure for the launch of NASA shuttles was available but had not been analysed (Vaughan, 1996).

The UnCoCoH model provides a focus for research and supporting practical action. Through the development of the model, we reconnect with the Carnegie School's interest in cognition, decision-making, and action in uncertain situations by transitioning cognition – individual perceptions of possible future events through to future-orientated action. It takes into consideration aspects of cognition by: recognising an individual's *perception* of situations, enabling the individual's *attention* to be directed at what is known and what might not be known, and encouraging the individual's *sensemaking* (Weick, 1979) and *sense-giving* (Gioia and Chittipeddi, 1991) through engaging in a process of *inquiry* (Schein, 2013; Garvin and Roberto, 2001). The transition to proactive action considers that, in all the decision situations that project leaders encounter, they will need to find ways to stabilise uncertainty, and one important way of doing this is through narratives (Kay and King, 2020; Winch and Sergeeva, 2022).

Narratives and narrating in construction project organising

One of the two processes of the action axis of the PLM is *projecting* – a term we take from Defoe (1697). The principal facets of *projecting* are narrating and storytelling (Winch et al., 2022). We focus here on narrating. Narratives play an enormously important role in *projecting* by connecting the present with the future, and are the essential means for maintaining or reproducing stability and promoting or resisting change in and around organisations (Vaara et al., 2016). They are, therefore, essential for decision-making under uncertainty (Kay and King, 2020). They are performative as words that do things (Austin, 1962) and therefore intendedly persuasive in nature. They are used by project leaders to convince stakeholders during project shaping and to mobilize resources during project delivery (Sergeeva and Winch, 2021).

When stabilising uncertainty, project leaders craft and communicate a project shaping narrative that inspires employees, excites partners, attracts customers and engages influencers and, perhaps most importantly, investors. The project shaping narrative is used to explain why the project exists and what makes it unique, the value and relationships it creates, and communicates these to both internal project team members and external stakeholders – in sum an image for the project (Sergeeva and Winch, 2021). The image-shaping narrative generates a project mission as a compelling 'why' statement for the project. Project narratives are communicated in spoken (e.g. talks, presentations), written (e.g. reports, business cases), and visual (e.g. videos, pictures, PowerPoint packs) forms to various internal and external stakeholders. The project narrative is also (re)iterated and restated in many different ways throughout the project life cycle to serve various purposes and audiences.

We can distinguish different types of narratives (e.g. project narrative about mission, innovation, sustainability, value creation), and each has important implications for project organising, shaping, delivery, and outcomes. Ante-narratives (Boje et al., 2016) are what come before a coherent and persuasive project narrative and, in effect, form alternative future possibilities of our world that can be created by *projecting*. Ante-narratives or 'before-narratives' are narratives that are not yet fully formed as project narratives and are still competing for the attention of stakeholders (Winch et al., 2022). Ante-narratives are often presented in speeches and talks, or published in reports, newspaper articles, and social media blogs as well as being the stuff of internal strategy debates within the organisation before the coherent reference narrative is formed about the project and the project mission becomes succinctly stated.

There are always counter-narratives to the project narrative, often mobilized by external stakeholders, to the dominant project narrative and ongoing interactions between them. Counter-narratives are 'the stories which people tell and live which offer resistance to, either implicitly or explicitly, dominant cultural narratives' (Andrews, 2004). The distinctive characteristics of counter-narratives are oppositional to the dominant project narrative. As demonstrated by Ninan and Sergeeva (2021) in the case of High Speed 2 (HS2), there are narratives of the need for a project; there are also counter-narratives that the project is not needed. The promoters are interested in supporting the completion of the megaproject, while protesters are interested in derailing the megaproject. They explored the role of labels in the sensemaking process through which these labels are maintained and contested in megaproject settings. The promoters labelled the HS2 megaproject as 'fast' and 'low-carbon'; the protesters labelled it as a 'vanity project' and as a project 'for the rich'. Focusing on counter-narratives enables us to capture some of the political, economic, social, and/or cultural complexities and tensions in projecting and the diversity of stakeholder positions in relation to the project narrative. The dynamic interaction between dominant and counter-narratives is part of the power game around project shaping.

Project identity is conveyed internally to the project team and the supply chain, whereas project image is projected to external stakeholders, such as investors, campaigners, and policy makers. Project identity narratives are about what project leaders tell the team in order to achieve shared understanding and vision; they are about a sense of what the delivery project organisation's purpose is that creates its 'DNA' (Ninan et al., 2019; Sergeeva and Winch, 2021). Project leaders communicate a narrative about project identity to the project team. This commitment is based on the connection of the group combined with the emotional value that is attributed to this connection. Project image narratives stimulate stakeholders to commit themselves to the project. Crafting a favourable image is import for gaining legitimacy and support from external stakeholders that in turn affect the delivery of project outputs. Projects require convincing narratives to build strong brand attributes and loyalty. This is why it is important to brand the project with a well-crafted external image from the start; hence, crafting a project image narrative as part of project shaping is essential for the successful delivery of projects from an external stakeholder management perspective (Winch et al., 2022).

Project leaders craft and maintain project narratives (about mission, scope, identity, image, innovation, sustainability, health, safety and wellbeing, value) throughout a project life cycle and their general work-life experience. They communicate and share their project narratives internally with the team and externally with people outside (through social media, Facebook, LinkedIn, Twitter). Sharing project narratives may have an impact on winning new projects, feeling proud of the work completed, and making new connections and contacts. There is an ongoing process of narrating and storytelling in different forms in construction project organising. Here we have shown how project narratives play a role in *projecting* as a way of stabilising uncertainty by connecting the present with the future.

Narratives and the process of narrating have important implications for the perceptions of the project. For instance, a negative narrative about a project crafted and communicated by external stakeholders is likely to have a damaging impact on the perceptions of the project held by the wider community. By the same token, a positive narrative about the project can give rise to favourable perceptions of the project held by the wider community. Hence, narratives and the process of narrating are important parts of transitioning from cognition to action in construction project organising.

Theoretical implications

Our concern for the relationship between cognition and action in construction project organising is rooted in the phenomenology of Schütz (1967) and his concept of 'projecting'. He argues – admittedly misconstruing Heidegger – that an action always has 'the nature of a project', and he uses Entwurf (construction drawings) as the noun (rather than Projekt; 1967, p. 59). So, it is appropriate to turn to the work of the New York School of relational sociology, which draws on Schütz to develop the concept of projectivity, for the broader theoretical implications for construction project organising of our argument in this chapter.

The New York School of relational sociology (Mische, 2011) is concerned with how people think about the future and the concept of *projectivity* (Emirbayer and Mische, 1998). Sociology has long been challenged in terms of understanding how people think about the future (Mische, 2014). By extension, we can suggest that this limitation also applies to those areas of management research that draw principally on sociology, such as organisation theory (Wenzel et al., 2020). Mische (2014) argues that this is because, in the Parsonian grand scheme of things, concerns of future orientation were left to economists, while sociologists got on with understanding the present. Economists then responded to this burden by providing us with EUT discussed above. However, we may add a more proximate reason rooted in the influence of Giddens on practice theory because he (Giddens, 1979) asserts a temporal conflation of the synchronic and diachronic (Saussure, 1959). In structuration theory, agency and structure are so tightly bonded in their mutual instantiation in space/time that the possibility of structure and agency evolving through different temporal rhythms is occluded, and 'temporal relations between structure and agency logically cannot be examined' (Archer, 1993, p. 70). The result is a timeframe that is entirely in the present, rather than the past or future.

The principal theoretical concern of Emirbayer and Mische (1998) is that human agency is fundamentally temporal, with a chordal triad of the iterative which reflects on the past, the projective which generates the future, and the evaluative in which experience is contextualized in the present. For researchers in construction project management, the likely concern is the projective, which has a dominant tone of 'narrative construction'. We suggest, therefore, that narratives are the principal way in which we imagine the future; hence, their crafting is an important aspect of temporal work (Kaplan and Orlikowski, 2013), defined as 'addressing tensions (implicitly or explicitly) among different understandings of the past, present, and future to settle on a strategic account for the organisation' (Shipp and Jansen, 2021, p. 334) – a perspective that aligns closely with Kay and King's (2020) from the perspective of economics. Projects are then the way in which those narratives are realized (Winch and Sergeeva, 2022).

A further perspective on temporality and the future is attention to the role of anticipation (Tavory and Eliasoph, 2013). According to this view, there are three levels of temporal experience. The first level is the 'protentions' that people experience in daily interactions. The second level is the 'trajectories' that people move through temporally as they make their way through the social world. These trajectories have two elements – narratives and projects. Narratives help people make sense of the future, while projects are how they move towards that anticipated future. The final level is the 'landscapes', which form the structured temporal elements of life, such as progression on an annual rhythm through the education system or the structure of the calendar itself.

Both Emirbayer and Mische (1998) and Tavory and Eliasoph (2013) draw heavily on Schütz (1967) in developing their arguments. Schütz develops an ontology that offers much insight for theorists of temporality in project organising. He argues that all purposive action, as opposed to reactive behaviour, has the nature of a protention or a vision of a completed future state which gives present meaning to that subsequent action that will bring forth that future state. Thus, while the protention is cognitive in that it exists as a perceived state, it is qualitatively different from a retention, which is inherently a perception about the past. However, because the protention, like a retention, is perceived as completed, 'the planned act has the temporal character of pastness' (Schütz, 1967, p. 61) and is therefore thought of in the future-perfect tense. We suggest, therefore, that projectivity is a potentially important concept for construction project organising research, and can be defined (Winch, 2023b) from a cognitive perspective as future-perfect thinking (Schütz, 1967) and from an action perspective as *projecting*. In turn, research on construction project organising can inform the projectivity research agenda, as the case of the Eden project shows (Winch and Sergeeva, 2022). We further suggest that this research agenda can most effectively be achieved by combining the New York School's concern with projectivity and a critical realist philosophy of science (Donati, 2018; Porpora, 2018).

Future research directions on projectivity in construction project organising

Some of the ideas set out in this chapter and elsewhere in this book may well prove useful, but we have also identified some areas where further research is clearly necessary. This constitutes a potential research agenda on projectivity in construction project organising. We highlighted the importance of researching project narratives and narrating in construction project organising research. Work has started on this agenda (Ninan and Sergeeva, 2021; Sergeeva, 2022; Sergeeva and Winch, 2021) and merits further investigation. From a projectivity perspective, narratives are about how future-perfect thinking is stabilized enough to mobilize the resources required for projecting. They are, therefore, core to leading construction project organising (Winch et al., 2022). Project narratives and narrating help shape the perceptions of a project from stakeholders, and hence have important implications for wider project organising aspects, such as organisational identity and image. Further research could explore the natures of different project narratives (e.g. about mission and vision, project outputs and outcomes) in project organising. Research could also start to identify the future visioning narratives in construction project organising (e.g. narratives about recovery from the pandemic, narratives about health and well-being, narratives about work–life balance). Through a deeper understanding of project narratives and narrating, we would better understand construction project organising and its future vision.

In addition, we must consider the role of the project owner. We have long understood that the owner (client) 'charters' the construction project (Boyd and Chinyio, 2006; Cherns and Bryant, 1984), and there is a growing body of research on the importance of owner project capabilities (Hui et al., 2008; Leiringer, 2023; Merrow, 2011; Morris and Hough, 1987; Winch and Leiringer, 2016), but much research remains to be done. From a projectivity perspective, cognition and action are linked within the owner domain (Winch, 2014) – that is to say, within the owner organisation and its broader socio-economic ecosystem. Present conceptualisations of strategic misrepresentation and optimism bias (Flyvbjerg et al., 2003) leave the development of these linkages as something of an empirical 'black box' that we need to get inside because, if the future is unknowable, then it logically follows that there is not some correct version of the future from which we are 'biased'.

Specifically, within this theme, the issue of the uses and abuses of cost–benefit analysis arises. Few would gainsay the contention that present CBA practice is 'broken' in some way (Flyvbjerg and Bester 2021; Self, 1970). The question is whether it can be fixed. The answer turns on whether one holds that uncertainty about the future is an epistemological or ontological problem. If it is the former, then the rigorous application of the technical fixes propounded by its advocates (Flyvbjerg and Bester, 2021; Sunstein, 2018) should be our aspiration; if it is the latter, then CBA can only be considered to be a very important sense-making tool supporting the feasibility criterion (Schütz, 1967) within future-perfect thinking.

These considerations link through to the growing concern in project organising research more generally around front-end definition (Babaei et al., 2021; Edkins et al., 2013; Williams et al., 2019; Zerjav et al., 2021), yet we still have little empirical research on the processes of project shaping that yields the insights obtainable from the kind of detailed empirical work offered by Hughes (1998) and Cusin and Passebois-Ducros (2015). Research methods drawing on ethnography, history, and political science would be particularly appropriate here to explore how cognition of a desired future becomes action initiating a project to achieve that desired future.

Conclusions

We began by summarizing the contribution of the Carnegie School to our understanding of management and organisations and indicated how their insights are still relevant today. Through this review, we have identified some opportunities for research in construction project organising. Our focus is on decision-making in organisations, by investigating the cognitive processes that influence decision-making and actions under conditions of uncertainty (Winch et al., 2022). We provided an overview of the MOC discipline, which provides the theoretical foundations for the exploration of uncertainty and project organising, which in turn has its foundations in the Carnegie School. We emphasized the importance of taking a cognitive perspective, as this places emphasis on individual processes that are reflected upon, learned, and socially constructed to provide an underpinning to actions associated with uncertainty and project organising. In developing our cognitive approach, we have built on Winch and Maytorena's (2011) cognitive approach to uncertainty on projects, and Kay and King's (2020) notion of 'radical uncertainty'. Our cognitive approach was developed by looking at how the cognitive framings shape organisational action. We consolidated relevant literature that attempted to understand risk and uncertainty and organisational action. The resulting UnCoCoH model links the cognitive approach with knowledge perspectives, leadership processes, and behavioural warning signs. This allowed us to reflect on the role of uncertainty in construction project organising and to identify the role of narratives in stabilising uncertainty through the project life cycle. Finally, we linked these concerns through to recent research in relational sociology on projectivity and suggested that a projectivity perspective on construction project organising combines cognition through future-perfect thinking and action through projecting (Winch, 2023b). We then indicated the sort of research agenda that this projectivity perspective suggests.

Our final comment is that the rather theoretical – and so inherently abstract – concerns explored in this chapter have important practical implications. The principal research contributions to construction project organising have taken a predominantly subjective perspective on time, seeing project organising as an emergent process (Winch and Sergeeva, 2022). While these theoretical developments have yielded many important empirical insights, they have lost one of the fundamental attributes of project organising – its goal-orientation towards a better future. We suggest that, as we live through the fourth industrial revolution and its net zero imperative (Winch, 2022), conceptualising projects as emergent processes is no longer adequate. The United Nations Sustainable Development Goals and the Paris Agreement are nothing if not goal-oriented. We therefore need to move beyond conceptualising projects as emergent processes and theoretically re-instate their inherent teleology. A projectivity perspective, we suggest, is one way to do this.

References

Ackermann, F., & Eden, C. (2011). Strategic management of stakeholders: Theory and practice. *Long Range Planning*, *44*(3), 179–196.
Alvarez, S. A., & Porac, J. (2020). Imagination, indeterminacy, and managerial choice at the limit of knowledge. *Academy of Management Review*, *45*(4), 735–744.
Ancona, D. G., Malone, T. W., Orlikowski, W. J., & Senge, P. M. (2007, February). In praise of the incomplete leader. *Harvard Business Review*, 92–100.
Andrews, M. (2004). Opening to the original contributions: Counter-narratives and the power to oppose. In M. G. W. Bamberg & M. Andrews (Eds.), *Considering counter-narratives: Narrating, resisting, making sense* (pp. 1–6). John Benjamins.
Archer, M. S. (1993). Taking time to link structure and agency. In H. Martins (Ed.), *Knowledge and passion: Essays in honour of John Rex* (pp. 154–173). I. B. Tauris.
Arikan, A. M., Arikan, I., & Koparan, I. (2020). Creation opportunities: Entrepreneurial curiosity, generative cognition, and Knightian uncertainty. *Academy of Management Review*, *45*(4), 808–824.
Aristotle. (1975). *Categories and de interpretatione*. Clarendon Press.
Austin, J. L. (1962). *How to do things with words*. Harvard University Press.
Aven, T., & Renn, O. (2009). On risk defined as an event where the outcome is uncertain. *Journal of Risk Research*, *12*(1), 1–11.
Babaei, A., Locatelli, G., & Sainati, T. (2021). What is wrong with the front-end of infrastructure megaprojects and how to fix it: A systematic literature review. *Project Leadership and Society*, *2*, 100032.
Barnes, M. (1983). How to allocate risks in construction contracts. *International Journal of Project Management*, *1*(1), 24–28.
Barnes, M. (1988). Construction project management. *International Journal of Project Management*, *6*(2), 69–79.
Bazerman, M. H., & Watkins, M. D. (2004). *Predictable surprises: The disasters you should have seen coming and how to prevent them*. Harvard Business School Press.
Boje, D. M., Haley, U. C. V., & Saylors, R. (2016). Antenarratives of organizational change: The microstoria of Burger King's storytelling in space, time and strategic context. *Human Relations*, *69*(2), 391–418.
Boyd, D., & Chinyio, E. (2006). *Understanding the construction client*. Blackwell.
Bryman, A., Bresnen, M., Beardsworth, A. D., Ford, J., & Keil, E. T. (1987). The concept of the temporary system: The case of the construction project. *Research in the Sociology of Organizations*, *5*, 253–283.
Bromiley, P., Koumakhov, R., Rousseau, D. M., & Starbuck, W. H. (2019). The challenges of March and Simon's organizations: Introduction to the special issue. *Journal of Management Studies*, *56*(8), 1517–1526.
Brown, A. D., Stacey, P., & Nandhakumar, J. (2008). Making sense of sensemaking narratives. *Human Relations*, *61*(8), 1035–1062.
Chapman, C., & Ward, S. (2011). *How to manage project opportunity and risk* (3rd ed.). Wiley.
Checkland, P. B. (2001). Soft systems methodology. In J. Rosenhead & J. Minger (Eds.), *Rational analysis for a problematic world revisited* (pp. 61–90). Wiley.
Cherns, A. B., & Bryant, D. T. (1984). Studying the client's role in construction management. *Construction Management and Economics*, *2*(2), 177–184.
Cleland, D. I., & King, W. R. (1968). *Systems analysis and project management*. McGraw-Hill.

Courtney, H., Kirkland, J., & Viguerie, P. (1997, Nov/Dec). Strategy under uncertainty. *Harvard Business Review*, 67–79.
Cusin, J., & Passebois-Ducros, J. (2015). Appropriate persistence in a project: The case of the wine culture and tourism centre in Bordeaux. *European Management Journal*, *33*(5), 341–353.
Cyert, R. M., & March, J. G. (1963). *A behavioral theory of the firm*. Prentice Hall.
Daniel, P. A., & Daniel, C. (2018). Complexity, uncertainty and mental models: From a paradigm of regulation to a paradigm of emergence in project management. *International Journal of Project Management*, *36*(1), 184–197.
Defoe, D. (1697). *An essay upon projects*. Cockerill.
DeMeyer, A., Loch, C. H., & Pich, M. T. (2002). Managing project uncertainty: From variation to chaos. *MIT Sloan Management Review*, *43*(2), 60–67.
Donati, P. (2018). An original relational sociology grounded in critical realism. In F. Dépelteau (Ed.), *The Palgrave handbook of relational sociology* (pp. 431–456). Springer.
Drouin, N., Sankaran, S., van Marrewijk, A., & Müller, R. (2021). *Megaproject leaders: Reflections on personal life stories*. Edward Elgar.
Eden, C., & Ackermann, F. (1998). Analyzing and comparing idiographic causal maps. In C. Eden & J. C. Spenders (Eds.), *Managerial and organizational cognition: Theory, methods and research* (pp. 192–209). SAGE.
Eden, C., & Spender, J. C. (Eds.). (1998). *Managerial and organizational cognition: Theory, methods and research*. SAGE.
Edkins, A., Geraldi, J., Morris, P., & Smith, A. (2013). Exploring the front-end of project management. *Engineering Project Organization Journal*, *3*(2), 71–85.
Edmondson, A. C., & Harvey, J.-F. (2017). *Extreme teaming: Lessons in complex, cross-sector leadership*. Emerald.
Emirbayer, M., & Mische, A. (1998). What is agency? *American Journal of Sociology*, *103*(4), 962–1023.
Flyvbjerg, B., & Bester, D. W. (2021). The cost–benefit fallacy: Why cost–benefit analysis is broken and how to fix it. *Journal of Benefit–Cost Analysis*, *12*(3), 395–419.
Flyvbjerg, B., Bruzelius, N., & Rothengatter, W. (2003). *Megaprojects and risk: An anatomy of ambition*. Cambridge University Press.
Galavan, R. J., & Sund, K. J. (2021). *Thinking about cognition*. Emerald.
Galavan, R. J., Sund, K. J., & Hodgkinson, G. P. (2017). *Methodological challenges and advances in managerial and organizational cognition*. Emerald.
Galbraith, J. R. (1977). *Organization design*. Addison-Wesley.
Garvin, D. A., & Roberto, M. A. (2001). What you don't know about making decisions. *Harvard Business Review*, *79*(8), 108–119.
Gavetti, G., Greve, H. R., Levinthal, D. A., & Ocasio, W. (2012). The behavioral theory of the firm: Assessment and prospects. *Academy of Management Annals*, *6*(1), 1–40.
Gavetti, G., Levinthal, D., & Ocasio, W. (2007). Neo-Carnegie: The Carnegie School's past, present, and reconstructing for the future. *Organization Science*, *18*, 523–536.
Giddens, A. (1979). *Central problems in social theory: Action, structure, and contradiction in social analysis*. Macmillan.
Gilovich, T., Griffin, D., & Kahneman, D. (2002). *Heuristics and biases: The psychology of intuitive judgement*. Cambridge University Press.
Gioia, D. A., & Chittipeddi, K. (1991). Sensemaking and sensegiving in strategic change initiation. *Strategic Management Journal*, *12*(6), 433–448.
Greenwood, J. D. (1999). Understanding the 'cognitive revolution' in psychology. *Journal of the History of the Behavioral Sciences*, *35*(1), 1–22.

Griffin, M. A., & Grote, G. (2020). When is more uncertainty better? A model of uncertainty regulation and effectiveness. *Academy of Management Review, 45*(4), 745–765.
Hobday, M., Davies, A., & Prencipe, A. (2005). Systems integration: A core capability of the modern corporation. *Industrial and Corporate Change, 14*(6), 1109–1143.
Huff, A. S., Milliken, F. J., Hodgkinson, G. P., Galavan, R. J., & Sund, K. J. (2016). A conversation on uncertainty in managerial and organizational cognition. In A. S. Huff, F. J. Milliken, G. P. Hodgkinson, R. J. Galavan, & K. J. Sund (Eds.), *Uncertainty and strategic decision-making* (pp. 1–31). Emerald.
Hughes, T. P. (1998). *Rescuing Prometheus: Four monumental projects that changed the modern world.* Vintage.
Hui, P. P., Davis-Blake, A., & Broschak, J. P. (2008). Managing interdependence: The effects of outsourcing structure on the performance of complex projects. *Decision Sciences, 39*(1), 5–31.
Judge, T. A., Bono, J. E., Ilies, R., & Gerhardt, M. W. (2002). Personality and leadership: A qualitative and quantitative review. *Journal of Applied Psychology, 87*(4), 765.
Kahneman, D., Sibony, O., & Sunstein, C. R. (2021). *Noise: A flaw in human judgment.* William Collins.
Kahneman, D., Slovic, P., & Tversky, A. (Eds.). (1982). *Judgement under uncertainty; heuristics and biases.* Cambridge University Press.
Kahneman, D., & Tversky, A. (1979). Prospect theory: An analysis of decision under risk. *Econometrica, 47*, 263–291.
Kaplan, S., & Orlikowski, W. J. (2013). Temporal work in strategy making. *Organization Science, 24*(4), 965–995.
Kay, J., & King, M. (2020). *Radical uncertainty decision-making for unknowable future.* Bridge Street Press.
Keynes, J. M. (1937). The general theory of employment. *Quarterly Journal of Economics, 51*(2), 209–223.
Klein, B., & Meckling, W. (1958). Application of operations research to development decisions. *Operations Research, 6*(3), 352–363.
Klein, G. (2007, September). Performing a project premortem. *Harvard Business Review*, 18–19.
Klein, G. (2017). *Sources of power: How people make decisions* (20th anniversary ed.). MIT Press.
Knight, F. H. (1921). R*isk, uncertainty and profit.* Houghton Mifflin.
Kurtz, C. F., & Snowden, D. J. (2003). The new dynamics of strategy: Sense-making in a complex and complicated world. *IBM Systems Journal, 42*(3), 462–483.
Lampert, C. M., Kim, M., & Polidoro Jr, F. (2020). Branching and anchoring: Complementary asset configurations in conditions of Knightian uncertainty. *Academy of Management Review, 45*(4), 847–868.
Lant, T. K., & Shapira, Z. (2001). *Organizational cognition: Computation and interpretation.* Lawrence Erlbaum Associates.
Leiringer, R. (2023) Owner project capabilities for large-scale infrastructure development projects. In G. M. Winch, M. Brunet, & D. Cao (Eds.), *Research handbook on complex project organizing* (pp. 301–11). Edward Elgar.
Loch, C. H., DeMeyer, A., & Pich, M. T. (2006). *Managing the unknown: A new approach to managing high uncertainty and risk in projects.* John Wiley & Sons.
Lundin, R. A., & Söderholm, A. (1995). A theory of the temporary organization. *Scandinavian Journal of Management, 11*(4), 437–455.
Maitlis, S. (2007). Triggers and enablers of sensegiving in organizations. *Academy of Management Journal, 50*(1), 57–84.

March, J. G., & Simon, H. (1958). *Organizations* (1st ed.). Wiley.
March, J. G., & Simon, H. (1993). *Organizations* (2nd ed.). Blackwell.
Mayer, J. D., Roberts, R. D., & Barsade, S. G. (2008). Human abilities: Emotional intelligence. *Annual Review of Psychology, 59*, 507–536.
Merrow, E. W. (2011). *Industrial megaprojects: Concepts, strategies, and practices for success*. Wiley.
Mische, A. (2011). Relational sociology, culture, and agency. In J. Scott & P.J. Carrington (Eds.), *The SAGE handbook of social network analysis* (pp. 80–97). SAGE.
Mische, A. (2014). Measuring futures in action: Projective grammars in the Rio+ 20 debates. *Theory and Society, 43*(3–4), 437–464.
Morris, P. W. G. (1973). An organisational analysis of project management in the building industry. *Build International, 6*, 595–615.
Morris, P. W. G. (1994). *The management of projects*. Thomas Telford.
Morris, P. W. G., & Hough, G. H. (1987). *The anatomy of major projects: A study of the reality of project management*. Wiley.
Müller, R., & Turner, J. R. (2010). *Project-oriented leadership*. Gower.
Neale, M. A., Tenbrunsel, A. E., Galvin, T., & Bazerman, M. H. (2006). A decision perspective on organizations: Social cognition, behavioural decision theory and the psychological links to micro and macro organizational behaviour. In S. T. Clegg, C. Hardy, T. B. Lawrence, & W. R. Nord (Eds.), *The SAGE handbook of organization studies* (2nd ed., pp. 485–519). SAGE.
Ninan, J., Clegg, S., & Mahalingam, A. (2019). Branding and governmentality for infrastructure megaprojects: The role of social media. *International Journal of Project Management, 37*(1), 59–72.
Ninan, J., & Sergeeva, N. (2021). Labyrinths of labels: Promoters' and protesters' narrative constructions in mega projects. *International Journal of Project Management, 39*(5), 496–506.
Oppenheimer, D. M., & Kelso, E. (2015). Information processing as a paradigm for decision-making. *Annual Review of Psychology, 66*, 277–294.
Packard, M. D., & Clark, B. B. (2020). Mitigating versus managing epistemic and aleatory uncertainty. *Academy of Management Review, 45*(4), 872–876.
Porpora, D. V. (2018). Critical realism as relational sociology. In F. Dépelteau (Ed.), *The Palgrave handbook of relational sociology* (pp. 413–429). Springer.
Pryke, S. (2020). *Successful construction supply chain management: Concepts and cases* (2nd ed.). Wiley Blackwell.
Puranam, P. (2018). *The microstructure of organizations*. Oxford University Press.
Ramasesh, R. V., & Browning, T. R. (2014). A conceptual framework for tackling knowable unknown unknowns in project management. *Journal of Operations Management, 32*(4), 190–204.
Rindova, V., & Courtney, H. (2020). To shape or adapt: Knowledge problems, epistemologies, and strategic postures under Knightian uncertainty. *Academy of Management Review, 45*(4), 787–807.
Saussure, F. D. (1959). *Course in general linguistics*. Peter Owen.
Savage, L. J. (1954). *Foundations of statistics*. John Wiley and Sons.
Schein, E. H. (2013). *Humble inquiry: The gentle art of asking instead of telling*. Berrett-Koehler Publishers.
Schoemaker, P. J. H. (1982). The expected utility model: Its variants, purposes, evidence and limitations. *Journal of Economic Literature, 20*, 529–563.
Schütz, A. (1967). *The phenomenology of the social world*. Northwestern University Press.

Self, P. (1970). Nonsense on stilts: Cost–benefit analysis and the Roskill Commission. *Political Quarterly, 41*(3), 249–260.
Sergeeva, N. (2022). Sustainability: Inclusive Storytelling to Aid Sustainable Development Goals. Princes Risborough: Association for Project Management.
Sergeeva, N., & Winch, G. M. (2021). Project Narratives that potentially perform and change the future. *Project Management Journal, 52*(3), 264–277.
Shipp, A. J., & Jansen, K. J. (2021). The 'other' time: A review of the subjective experience of time in organizations. *Academy of Management Annals, 15*(1), 299–334.
Snowden, D. J., & Boone, M. (2007). A leader's framework for decision-making. *Harvard Business Review, 85*(11), 69–76.
Staw, B. M., Sandelands, L.E . & Dutton, J.E . (1981). Threat–rigidity effects in organizational behavior: A multilevel analysis. *Administrative Science Quarterly, 26*, 501–524.
Stephens, P. (2003, December 12). The unwitting wisdom of Rumsfeld's unknowns. *Financial Times.*
Stoker, B. (1897). *Dracula.* Hutchinson.
Sunstein, C. R. (2018). *The cost–benefit revolution.* MIT Press.
Taleb, N. N. (2008). *Black swan: The impact of the highly improbable.* Penguin.
Tavory, I., & Eliasoph, N. (2013). Coordinating futures: Toward a theory of anticipation. *American Journal of Sociology, 118*(4), 908–942.
Vaara, E., Sonenshein, S., & Boje, D. (2016). Narratives as sources of stability and change in organizations: Approaches and directions for future research. *Academy of Management Annals, 10*(1), 495–560.
Vaughan, D. (1996). *The challenger launch decision: Risky technology, culture and deviance at NASA.* Chicago University Press.
Von Neumann, J., & Morgenstern, O. (1944). *Theory of games and economic behavior.* Princeton University Press.
Walsh, P. (1995). Managerial and organizational cognition: Notes from a trip down memory lane. *Organization Science, 6*(3), 280–342.
Weick, K. E. (1979). *The social psychology of organizing* (2nd ed.). Addison-Wesley.
Weick, K. E. (1995). *Sensemaking in organizations.* SAGE.
Weick, K. E., Sutcliffe, K. M., & Obstfeld, D. (2005). Organizing and the process of sensemaking. *Organization Science, 16*(4), 409–421.
Wenzel, M., Krämer, H., Koch, J., & Reckwitz, A. (2020). Future and organization studies: On the rediscovery of a problematic temporal category in organizations. *Organization Studies, 41*(10), 1441–1455.
Wilden, R., Hohberger, J., Devinney, T. M., & Lumineau, F. (2019). 60 years of March and Simon's organizations: An empirical examination of its impact and influence on subsequent research. *Journal of Management Studies, 56*(8), 1570–1604.
Williams, T., Vo, H., Samset, K., & Edkins, A. (2019). The front-end of projects: A systematic literature review and structuring. *Production Planning & Control, 30*(14), 1137–1169.
Williamson, O. E. (1975). *Markets and hierarchies.* Free Press.
Winch, G. M. (2001). Governing the project process: A conceptual framework. *Construction Management & Economics, 19*(8), 799–808.
Winch, G. M. (2010). *Managing construction projects* (2nd ed.). Wiley-Blackwell.
Winch, G. M. (2014). Three domains of project organising. *International Journal of Project Management, 32*(5) 721–731.
Winch, G. M. (2015). Project organizing as a problem in information. *Construction Management and Economics, 33*(2), 106–116.

Winch, G. M. (2022). Projecting for sustainability transitions: Advancing the contribution of Peter Morris. *Engineering Project Organization Journal, 11*(2). https://doi.org/10.25219/epoj.2022.00101
Winch, G. M. (2023a) Uncertainty. In G. M. Winch, M. Brunet, & D. Cao (Eds.), *Research handbook on complex project organizing* (pp. 17–25). Edward Elgar.
Winch, G. M. (2023b). Projectivity. In G. M. Winch, M. Brunet, & D. Cao (Eds.), *Research handbook on complex project organizing* (pp. 36–45). Edward Elgar.
Winch, G. M., Brunet, M., & Cao, D. (2023). Introduction to the Res*earch Handbook on Complex Project Organizing.* In G. M. Winch, M. Brunet, & D. Cao (Eds.), *Research handbook on complex project organizing* (pp. 1–10). Edward Elgar.
Winch, G. M., & Leiringer, R. (2016). Owner project capabilities for infrastructure development: A review and development of the 'Strong Owner' concept. *International Journal of Project Management, 34*(2), 271–281.
Winch, G. M., & Maytorena, E. (2011). Managing risk and uncertainty on projects: A cognitive approach. In P. W. G. Morris, J. K. Pinto, & J. Søderlund (Eds.), *The Oxford handbook of project management* (pp. 345–364). Oxford University Press.
Winch, G. M., Maytorena, E., & Sergeeva, N. (2022). *Strategic project organizing.* Oxford University Press.
Winch, G. M., & Sergeeva, N. (2022). Temporal structuring in project organizing: A narrative perspective. *International Journal of Project Management, 40*(1), 40–51.
Winter, M., & Szczepanek, T. (2009). *Images of projects.* Gower.
Zerjav, V., McArthur, J., & Edkins, A. (2021). The multiplicity of value in the front-end of projects: The case of London transportation infrastructure. *International Journal of Project Management, 39*(5), 507–519.

8 Construction safety management: the case for a new approach to research-informed change

Steve Rowlinson

The construction industry as an arena for Occupational Health Safety and Wellbeing

The construction industry has the poorest safety performance of any industry worldwide, despite many decades of research that has tried to address its failings. Why is this the case? Is it a failure in terms of the legislation regarding safety at work? Is it a failure of safety management systems that are in place but do not function effectively? Or is it the case that research into construction project safety management has been ill-conceived, misdirected, and even irrelevant to practice? What is known is that the construction industry is characterised by a whole range of issues around its fragmentation, leading to communication problems, conflicts, and a lack of collaboration. This begins with the client that often lacks a long-term view of the management of its assets, and is exacerbated by poor organisation on site with the multiple layers of subcontracting and the lack of proper, complete design of permanent and temporary works. These issues beset the industry not only in terms of safety, but also in terms of productivity, quality, and sustainability. The result is that the industry regards itself as unique, and the problems it faces are seen as being wicked in nature. This is probably not the case; the real issue is that the industry does not speak with one voice and rarely takes cognisance of the research and information available outside of the fields of engineering management, project management, and lean production. Given the intensely competitive bidding systems that beset the industry, Occupational Health Safety and Wellbeing (OHSW) knowledge is rarely shared, as past OHSW performance is part of the adjudication system in these outdated bidding systems. The winner's curse appears time and again due to the competitive bidding system and the boom–bust cycles of construction demand. Consequently, many of construction's best people are forced into being narrow-thinking followers of the production versus quality versus safety trade-off. The multiple challenges that this creates for safe production can only be changed through education at tertiary and professional levels, as well as through informed research and culture change at corporate and project levels.

A peculiarity of the construction industry is that design, production, and use are divided in domains of different professional and commercial organisations. This fragmentation takes place at macro and micro levels. In terms of its institutions and governance, construction is a very different prospect from manufacturing and other industries. There is a very high degree of subcontracting throughout the project process. Effective legislation and management have been hindered by this fragmentation and by the fact that the industry builds a wide range of products. In terms of safety management, therefore, one size does not fit all. The design, construction, commissioning, and facilities management process is subject to a whole series of institutional constraints, for instance, building approvals, fire safety approvals, and town planning restrictions, to name but a few. Hence, conducting and generalising from construction research is a task fraught with difficulty.

The construction industry differs from manufacturing industries in terms of its characteristics and business models (see Hillebrandt, 2000, and Hillebrandt & Cannon, 1989). To understand the differences and how organisation affects the way business and production are undertaken, the seminal work of Joan Woodward (1965) and her contingency analysis of different organisational forms for different industries are still relevant. Woodward argued that 'core technologies' determine an organisation's methods. Much of the early research into OHSW was undertaken in manufacturing businesses and the military, both of which have a much more stable organisational context than construction. As such, applying these findings requires some careful thought as to how they fit into the context of the construction industry and how to make use of extant theory.

Drilling down into the specifics of the construction industry's operating context, there is a clear divide between head office personnel (corporate) and their functions, and the individual projects of the business. The process of tendering relies heavily on the estimating team producing a competitive bid. This does not necessarily involve the project managers and site agents. Consequently, the project manager must complete on time and to budget, but has little or no say in putting the bid together. Often, the project manager exhibits directive leadership, demanding that engineers, foremen, and workers meet deadlines and budgets that are often unachievable. This 'bulldozer' approach knocks on to the subcontractors and others, who are often paid late, or not paid at all, for work that they have completed. This produces a whole range of tensions and paradoxical demands between corporate and project business in relation to workforce, safety, quality, and sustainability, with each project being run as almost a new business.

In attempting to address this complex operating context, early work on construction safety published in construction-related journals was practice-based, focusing on the description and deconstruction of approaches to safety management rather than looking for underlying theories and constructs to highlight how OHSW is mismanaged in the industry. In this chapter, the dominant approaches to understanding safety and health concerns is critically reviewed. This provides a picture of how business has addressed safety issues over time, but also reveals how construction management research (CMR) has had little impact on OHSW performance, despite changes in regulation. By looking outside of the sector to the ways in which OHSW has been tackled elsewhere, it is argued that a more effective and impactful research agenda might result that can help to address what is perhaps the industry's greatest failing.

A brief history of OSHW research and theory

In this section, approaches taken to researching and informing OSHW practice are reviewed as a backdrop to setting out a more robust agenda for construction safety research.

Accident causation theories

The seminal work of Herbert W. Heinrich (1931) and his domino theory was the first scientific approach to accident prevention. His idea was straightforward: to prevent accidents, the cause of accidents must be clearly described. His domino sequence presented a chain reaction, with each of the factors dependent on the preceding factor. Heinrich postulated that an injury only occurs as the result of an accident, which is a result of either personal or mechanical hazards. These occur through the fault of careless persons or due to poorly designed or improperly maintained equipment, and faults of persons can be innate or acquired because of social environment or genetics.

In investigating accidents, Heinrich was concerned with their proximate causes, emphasising that the accident should be the focus of investigation whether or not injury or property damage occurs. He set out the process of recording incidents, not just accidents, arguing for investigating all mishaps, supposing that eliminating non-injury incidents facilitated eliminating minor and serious injuries. This approach of investigating near-misses preceded current construction practice by decades, and was underpinned by his accident triangle (commonly presented as his 'law'): for every one accident there are 30 incidents and over 300 unreported occurrences.

Heinrich (1931) also set out a corrective accident sequence, the 3Es, which involved *engineering* as a first line of defence against hazards through product design or process changes. *Education*, not just training, should be given to all workers so they understand safety and its processes; it is management's duty to ensure education is both provided and effective. Third, *enforcement* of rules, regulations, and standard procedures should be overseen by management. The domino sequence is often wrongly interpreted to be based on human errors as the root cause of accidents. However, while Heinrich believed that overload, physical, and psychological stress at work were important causes, he also recognised the importance of environmental and situational factors. He further recognised the conflicting goals and struggles between workers and management and, by extension, the stresses that workers and project managers experience due to their different perspectives and priorities. This recognition stimulated the broadening of research into human factors, systems thinking, risk analysis, and reliability of systems.

Theories of accident causation developed as manufacturing and service industries grew worldwide. A significant contribution came with human factors theory, which attributed accidents to chains of events that were caused by human error. Factors leading to human error are: physical or mental overload; inappropriate responses to changing conditions; and inappropriate activities that the worker was not trained for or does not understand. Bird (1966) provided a protocol for applying this theory. For decades, researchers tinkered with this concept. For example, Petersen (2003) extended it by expanding the concepts of overload (physical, physiological, or psychological), decision to err (deliberately or by circumstance-violating procedures), and traps (incompatible physique, design of the workplace, or culture of the organisation – behaviours that are encouraged or discouraged). This evolving approach was often termed accident/incident theory and moved thinking in the direction of systems theory, which saw safety from three perspectives: person (host), machine (agency), and workplace (environment). The understanding of human error was explored further through epidemiology theory, which investigated workers' predispositions for choosing certain actions, and situational characteristics, such as peer pressure and poor attitudes towards safety and risk taking, as causes of accidents (e.g. Robertson, 1992).

Concurrent with the development of human factors theory, significant research by Eric Trist at the Tavistock Institute for Social Research utilised operations research analysis from a socio-technical systems perspective. Trist had studied longwall coal mining in the Nottinghamshire collieries and wrote a seminal article on the psychological consequences of the method on the miners (Trist and Bamforth, 1951). In discussing work groups, Trist noted, 'Groups ... were free to set their own targets ... production could be adjusted to the age and stamina of the individuals' (p. 7), and 'unfavourable and changing conditions are encountered at the coal face which are impossible to predict or alter' (p. 20). Both observations could have been made of construction sites. Hence, it is not surprising that the Tavistock Institute became involved in research in the construction industry, following Banwell's (1964) report on the placing and management of building contracts. This was an important departure for construction research as the industry invited outsiders to study it and operational researchers and social scientists conducted the research. Important studies were undertaken by Bryant and others on the role of the client in building, and Cherns and Bryant (1984) returned to this theme over 20 years later.

In the 1960s, the power generation and aviation industries started to take on approaches based on engineering and statistical concepts, such as reliability engineering, to quantify probabilities of failure and to focus on tasks with the highest risk. Risk identification analysis is crucial to safety management. It can be undertaken at the corporate, project, work section, or individual worker levels. It is multi-faceted and can be deep-seated within organisations. There has been a strong focus on this in construction management practice, but risk assessments have become generic, systematised, and text based, which has reduced their effectiveness. Beyond reliability engineering, systems analysis – for example, fault tree analysis (also developed in the 1960s) – focused on eliminating and controlling hazards. Interactions among components, not just component failures, were investigated, including non-technical aspects of systems, such as working patterns, location, and general environmental issues. In this method, operational decision-making is assessed by working backwards through a decision tree and conducting possible incident analyses. The assumption is that incidents arise from interactions between people, machines, and the environment. The chain of events is not a simple linear causality, but a complex system of different causal connections. Systems models view accidents as a control problem, and consider failure the result of a lack of system safety constraints. The intention is not to produce a 'blame' model but an 'understanding why' model. These 'engineering solutions' to safety management, although not neglecting the principles laid out by Heinrich, were applied in a deterministic fashion in many sectors, and the underlying viewpoint of managers was that risk – and by extension, accidents – can be engineered out of production processes.

Haddon (1973) took a different view to the dominant orthodoxies outlined above. He assumed that accidents were inevitable, so minimising their impact was the goal. He put forward the energy release theory, addressing accidents in terms of reducing and removing energy released during an incident. Haddon used this theory to propose seatbelt regulations for cars, as well as energy-absorbing steering columns and penetration-resistant windshields, providing a barrier to energy transfer, not necessarily stopping the incident. His thinking, in terms of the inevitability of accidents, was shared by others who came up with solutions that were not focused purely on damage minimisation but attitudinal and systemic change.

Contemporary views

In more recent times, scholars have set about rationalising what is wrong with safety management. As the new millennium approached, they investigated improvements by looking at OHSW from different perspectives, questioning established thinking. The inevitability of accidents became accepted, rather than the idea that zero harm could actually be achieved. Accepting the inevitability of accidents enabled alternative thinking and approaches to OHSW that lessened their impact.

James Reason (2000a) proposed the Swiss cheese model, explaining it as a trajectory of accident occurrence: many layers of defence lie between hazards and accidents; there are flaws in each layer that, if aligned, lead to accident occurrence. Factors that contribute to accidents are organisational/systems issues, the local workplace, and unsafe acts. Reason identified two common forms of unsafe acts, namely, rule-based mistakes and knowledge-based mistakes. This model is useful and simple when used correctly. Reason (2000b) identified four safety paradoxes:

> safety is defined and measured by its absence rather than its presence; defences such as barriers and safeguards protect a system but can also cause the system to break down; in attempting to minimise error, organisations limit variations in workers actions yet variability and adjustments for unexpected events actually maintain safety in a dynamic environment and stimulates learning; the belief in the attainability of zero accidents seriously impedes the realisation of safety goals. (p. 4)

In addressing the issue of blame-seeking, Reason and Hobbs (2003) introduced the concept of error management and proposed 12 principles of workplace management, shaping a different way of approaching accident analysis. They stated that errors are not intrinsically bad, and the best people can make the worst mistakes. Error management is about managing the manageable, making good people excellent, and aims at continuous reform rather than local fixes. There is no one best way:

> It is simply not possible to order in a package of Error Management measures, implement them and then expect them to work without further attention. You cannot put them in place and then tick them off as another job completed. In an important sense, the process – the continuous striving toward system reform – *is* the product. (Reason and Hobbes, 2003, p. 101, emphasis in original)

Reason's simple approach to error management consists of three activities: reduction of errors, containment of the tendency to err, and managing these so that the intervention remains effective. The three principles behind this are described as: learning culture; just culture; and reporting culture. Learning culture emphasises learning from incidents and acting on what has been learnt. The concept of a just culture involves building trust between all of those in the organisation in that they are free to report incidents and accidents, without blame, with the goal of improving safety. However, a just culture is not a no-blame culture. There are certain actions that are classed as contraventions of the overarching organisation culture, and the line between acceptable and unacceptable behaviour is clearly drawn and understood by all. However, in most construction companies, there is a tendency to pick out one element of a system or theory, in this case just culture, and apply it haphazardly rather than taking on board the whole system. This is a recurrent fault in construction firms in using research outputs.

Behaviour-based safety

In 1980, Sulzer-Azaroff presented the theory of behaviour-based safety (BBS) and introduced seven basic principles to reinforce appropriate (safe) behaviour through rewards and recognition (Sulzer-Azaroff, 1980). These were adopted by Cooper and colleagues in research on construction firms in the UK, and are as follows: identify an intervention target; identify factors or activities that need addressing; provide motivation to behave in the desired manner; reward positive outcomes of desired behaviour; document information to form an action plan; plan and make interventions; review and enhance; and reward successful interventions (Cooper et al., 1994).

A problem with BBS is finding interventions that workers can buy into. In many attempts at BBS, the underlying, basic safety infrastructure did not exist. Without this underlying minimum level of management preparation, BBS was doomed to fail. A typical example of this can be found in the work of Lingard and Rowlinson (1998), where a contractor bought into this new approach to safety improvement, but relied on the researchers and safety department to implement an intervention that was not supported by the basic physical safety infrastructure and management commitment, as the interventions were 'pushed' onto the project manager from a corporate director.

Accidents are inevitable?

Despite the research of Reason and others and the successful application of their approaches to safety management in other industries, there is a continuing focus in construction on blame culture and human error. Company directors are perplexed by the fact that although ISO (International Organization for Standardization)-accredited safety management systems are in place, accidents still happen. There is an assumption that if there is a system in place and an accident occurs, someone must have broken the rules and not followed the system. This assumption is addressed and critiqued in this section.

Dekker (2003) addressed the practice of blaming occupational mishaps and industrial accidents on 'human error'. He saw human error as a deeply problematic aspect of performance, if indeed it is a separate category of performance at all. Dekker's view was that human error was an effect rather than a cause. The cause lay deeper inside the systems implemented. This emanated from Dekker's 'new view' in ergonomics. He saw human error as 'a judgement made in hindsight', while what workers do makes sense to them at the time. He put forward the view that accidents are emergent phenomena without clear root causes and that, in business, deviance from procedures is generally accepted as normal, so organisations have 'messy interiors' no matter what systems they have in place. He further argued against human error as an explanatory factor, as it is used in many different ways: sometimes a judgement; sometimes a cause; sometimes a process; and sometimes an effect. Similar views have been expressed by, for example, Amalberti (2001) and Woods and Cook (2002).

Dekker stated that the old view saw human error as the cause of accidents, with engineers and managers believing that the systems they have devised are safe, if only workers would follow the processes and procedures. His 'new view', in contrast, takes the stance that human error is not an explanation but, rather, demands an explanation, as it may be an effect of failure in the system. Systems are not inherently safe, as they often need to meet multiple goals and people have differing perceptions of these systems. The starting point of the new view is the local rationality principle (cf. Simon, 1969) – that is, people do not come to work to do a bad job. What people do and decide must make sense to them given their knowledge, perspective, and understanding of the situation at the time, otherwise they would not do it. Thus, he claims that the search for the cause of failure is illusory; there is no such thing as the cause of an accident: 'trying to find out *the* cause of an accident is just as bizarre as trying to find out *the* cause of not having an accident' (Dekker, 2003, p. 215, emphasis in original).

Putting these ideas into context and critiquing the search for blame, Hollnagel (2006) provides a critical review of accident analysis and human error. He argues that, in the 1950s, three issues were used to define the root causes of accidents: technical failures, human error, and 'others'. He then shows how ergonomic approaches to accident analysis and machine interactions were later recognised as important underlying causes, followed by an increased focus on information processing models whereby the cause of accidents was put down to complexities in the system that the human mind could not deal with. In reviewing cognitive systems engineering, he notes how it focuses on how human performance is influenced by context. In his view, the fallacy of the socio-technical approaches, by focusing on context, perhaps misses out on information processing approach issues. He concludes that human error is not a specific category of accident causation. By going back through the whole process and identifying the mismatches between the human and the technical systems, processes of 'human error' can be avoided. This is a contentious but thought-provoking argument that aligns with the findings of Trist and the Tavistock Institute.

High-reliability organisations

The above discussion of different viewpoints and schools of thought assists our understanding of high-reliability organisations (HROs), or organisations operating in risky environments that have no catastrophic accidents. HROs emphasise continuous questioning and change of procedures and processes in order to not fall into the trap of accepting destructive deviance or deliberate bending of the rules to bad effect. They are founded on the concept of a just culture whereby all incidents can be reported without fear of management sanctions, and all function on the basis that they enable and expect authority to be simultaneously centralised and decentralised (Hopkins, 2007). In 'A Practical Guide to Becoming a High-Reliability Organisation', Hopkins (2021) goes through a wide-ranging discussion of HRO and gives a brief description of its origin in the 1980s at the University of California, Berkeley. He discusses the definition of HROs, starting with an extensive list of bad news situations that are generally not reported. For an organisation to be an HRO, these 'bad news' situations must be reported within a just culture. For example, bad news may be identifying procedures and rules that are not appropriate or are too complex to follow, influencing the quality of work. This list could have come from any construction site. Indeed, all these situations were reported on the sites of a major construction contractor in Hong Kong in a piece of (unpublished) safety climate contract research by the author. Hopkins finishes this enlightening paper by citing Weick et al. (1999) and the shift from an HRO to a mindful organisation. Weick and Sutcliffe (2001) describe the five processes of mindfulness by which HROs manage the unexpected. HROs: monitor 'small failures'; are 'reluctant to accept simplification'; remain 'sensitive to operations'; develop and maintain 'a commitment to resilience'; and practice 'deference to expertise'. Together these five processes produce a collective state of mindfulness. Hopkins concludes that defining an HRO is difficult and is better described through the five concepts of mindfulness. HROs are identified by performance and no major incidents in diverse, high-risk businesses, so they are most easily described by their traits rather than their structures, processes, or procedures.

The construction industry is well known for its blame culture, whether it be in relation to OHSW, cost overruns, or project delays. This ingrained attitude of looking for a culprit exists at all levels in the industry. It is time that the industry matured and took on board the concepts of HRO, just culture, and respect for the workforce. This attitude change will not be instant, given the attitudes of hard-nosed project managers, but will involve a long haul of gradual change, in the way that LGBTQ individuals achieved fair and proper treatment in society.

Construction management safety research

The discussion up to this point provides a sound foundation for understanding the management of safety within organisations and provides theoretical frameworks within which OHSW can be researched. However, most of this theorising has been done in more static environments of manufacturing, aviation, and energy, rather than the fluid and change-prone domain of construction. Adapting such frameworks has proven difficult for construction firms to implement, as there is a tendency for project managers and directors to impose a quick-fix version of what they have learnt, appoint a safety manager to 'run the system', and return focus to productivity and cost saving.

In this section, contemporary approaches by the CMR community to investigating OHSW in the construction industry are presented. Specifically, we move on from accident analysis and prevention and in the direction of active safety management and intervention. A range of research topics and approaches are reviewed, and some of the key researchers in the field are identified. We provide a view of how workers manage their safety by developing a workgroup safety climate, and how embedded issues, such as progress pressure (PrP) and reward systems, affect this. Other research areas, such as leadership and business practices, are discussed. Mental health issues and accident analysis research are also introduced.

Safety climate

Zohar (1980, 2000) defines a safety climate as a set of shared perceptions among workers concerning procedures, practices, and behaviours that are supported and rewarded to ensure safety when high-risk operations are being performed. However, instances occur when behaviours do not match the espoused climate (i.e. the safety culture of the organisation is weak, and other objectives, such as productivity, take precedence). For example, Walker (2010) looked at a workgroup that developed a counterculture to a positive safety climate. Taking an ethnographic approach, he uncovered the dilemma that work gangs faced in working safely and maintaining production in what he describes as a 'pathological organisation' that does not reward participation or actively communicate with its workforce. He describes the group as socially constructing danger, injury, and safety for themselves. He interprets the safety climate as a sham that has been contrived by management, making the workgroup constructed as a functioning counterculture. This is an interesting, and realistic, example. which is validated by the ethnographic nature of the research. By presenting examples of the contradictions between the understanding of management and the practical implementation of safety, involving risk taking that is carefully calculated by the work group, the conflicts between management and the work group over productivity and safety are clearly presented.

Oswald et al. (2017) looked at the rewards system for health and safety performance on a construction site. Money and vouchers were popular rewards, while certificates and public recognition were unpopular. As a slap in the face to tokenism and a reflection of one aspect of a weak safety climate, awardees did not believe that they deserved an award for 'just doing their job'. Perceived differences in the value of prizes caused resentment as 'bosses' received more. This showed a failure to develop a strong safety climate, with workers not buying in to petty rewards and seeing management commitment as being disingenuous. In an interesting counterpoint to this article, Oswald et al. (2018) provide a prescient description of compensation culture, and unjustified, frivolous, or fraudulent legal claims being made following safety accidents, conditions that are prevalent in the UK construction industry. This research was part of a three-year, multi-faceted ethnographic programme on a large construction project in the UK. Fraudulent compensation claims against employers are socially acceptable behaviour in many industries, but are particularly prevalent in the construction industry. More worryingly, the findings show how many organisations recognised that this happened and tried to protect themselves from potential liability. The emphasis moved from management of safety to management of potential compensation claims. The authors claim that the traditionally dominant positivist approach to research has overlooked socio-cultural aspects of the workgroup and the business aspects that influence the safety climate.

The above examples highlight a failure to engage workers in terms of developing a sustainable safety climate within workgroups, construction projects, and the organisation as a whole. The safety climate should be visible through the mindfulness displayed at all levels in the organisation. However, this failure to engage is not just present in OHSW but also in employment and payment issues, work pace pressure, skills training, and quality concerns. Thus, the conditions of employment in the industry and the separation between corporate decision-making and project management are the milieu within which competing and paradoxical demands are placed on employees. To summarise, the construction industry performs poorly in terms of worker engagement, and a different perspective is required to explore the root causes of this situation, as the underlying causes of weak safety climate and culture continue.

Leadership

A sustainable safety culture needs leadership at both the corporate and project levels. The adage 'walk the walk' holds true on construction sites where 'savvy navvies' see through half-hearted attempts to promote safety awareness and compliance where production takes precedence over safety. Lingard et al. (2019a) looked at leadership communication and its effect on safety climate and workgroup behaviour. Supervisors' transformational leadership strongly predicted OHSW participation, showing genuine commitment to change. However, OHSW compliance was developed through transactional leadership, contingent rewards, as well as supervisors' communication practices. The authors recommended leveraging these relationships through a supervisory leadership development programme. On the same theme, Oswald and Lingard (2019) developed a health and safety leadership maturity model. They identified six key areas as having an impact on performance in the form of relationships between: foreman and subcontractor supervisor; foreman and frontline workers; subcontractor supervisor and supervisor; frontline leaders and OHSW advisors; leadership styles of the foreman; and supervisor and inter- and intra-workgroup communication.

Business practice

In the same ethnographic study described above, Oswald et al. (2020) explored several issues including competitive bidding practices and their impact on corporate and project management. The research focused on perverse forms of tendering whereby the successful bid is the lowest – at times referred to as 'the winner's curse'. Consequently, cost saving is implemented to recoup the potential profit lost in the bidding process. Safety risks increase through cheaper and poor-quality equipment, machinery, and temporary structures.

Following an inductive approach, Lingard et al. (2019b) researched improvement in infrastructure projects in Australia, focusing on the Victorian Government New Public Management Approach to commercial management of infrastructure projects. They found that industry did not buy in to the government initiative and operated under 'business as usual'. There was a clear disconnect between what the client intended with the new management approach and the managerialist, or productivity (PrP), approach of the contractors and consultants, which focused on speedy, profitable delivery. The authors highlighted the importance of understanding the different and competing institutional logics driving the practices of the involved parties, which is something that has been overlooked in construction OHSW research until recently.

The above illustrates that conflicts between production, quality, and safety demands – the iron triangle – exist throughout the construction process, from bidding, subcontractor selection and management, and project leadership to production – and not just in the domain of OSHW. Decision-making is top-down, without discussion, and is a root cause of the inefficiencies of construction project management in general. This disconnect is a major factor in the range of project management failures that exist in the industry.

Mental health

Mental health is a major issue for workers and managers in the construction industry. The construction industries in Australia and the UK report exceedingly high suicide rates compared with other industries. Oswald et al. (2019) identified 15 studies worldwide that indicated mental health as a significant problem in construction. Patching (2022), one of the project managers for the Sydney Olympics, researched stress in construction project managers and concluded that they were mentally stressed due to the way projects are run, as well as work–life balance issues. He proposed that both the nature of project management and corporate decision-making need to change to relieve pressures of stress, coping, and harm, which are rarely addressed by businesses yet are central to the life of a project manager. Bidding decisions, which set the tone for projects, are made by a select group. The conflicts between production and OHSW arise on site but are driven by directions from the head office and personal achievement needs in a transient industry that has an unstable workload. This is a moral and ethical issue that needs to be highlighted with sound, multi-disciplinary research from psychology, sociology, and cultural studies, and embedded in a longitudinal research framework involving interventions and structural change. The stress at work and home brought about by long hours in confrontational meetings and with rapid decision-making requires ethnographic research. Furthermore, there is a need to employ theoretical constructs suited to the nature of the problem. Construction management researchers in general have taken an overly simplistic viewpoint in formulating research; thus, many outcomes have been piecemeal and unconvincing to the industry.

Accident analysis, prediction, and hazards

Heinrich studied the sources and causes of accidents, rationalised their causes, and proposed actions involving machines and equipment to make the workplace safe. Reason and others took this further and viewed OHSW through a socio-technical systems lens (as in Trist's seminal work in the Nottinghamshire coal mines), taking systems, environment, and business culture into account. Notable CMR contributors to this area of research were Gibb et al. (2006), who developed the ConCAS approach to accident analysis. This was extended further in a piece of abductive research by Harvey et al. (2018).

Nonetheless, CMR in general mostly lacks a sound theoretical basis, or a context in which it can be implemented. Hallowell et al. (2013) posited that safety practices should be measured during construction to trigger change before injury occurs. Tixier et al. (2016a, 2016b) attempted to build a database of accidents using artificial intelligence to analyse construction accidents and predict their severity. However, prediction can only be as good as the data gathered and requires mindful implementation. Furthermore, with current attitudes in construction, recommendations from such analyses are unlikely to be implemented because of the PrP-driven approaches of both the client and project manager. A major issue is that accident and incident reports are often used for compensation claims and in prosecutions for breach of safety regulations. Thus, wholesale analysis is unlikely to reveal anything new about accident causation and prevention. Hallowell has since partnered with major businesses as a mechanism for not just conducting research but also having findings implemented. This is a laudable approach to making research relevant to industry by providing a platform for learnings to be disseminated and absorbed into industry culture, but it is rarely done.

Much research has focused on identifying and recognising hazards. Having recognised a hazard, workers need to be able to identify the danger that it presents and manage it. Bhandari et al. (2020) devised an agenda and rubric for training in hazard recognition and mindfulness, a key skill needed in HROs. Hazard recognition by adopting virtual reality training has been researched (e.g. Albert et al., 2014). Hinze (1997) proposed a distraction theory assuming that workers are more likely to experience an incident or accident when they are distracted. Research has been conducted in barometric chambers and on sites in studies on human subjects (Rowlinson and Jia, 2013) where physical factors, such as blood pressure, heart rate, and sweat rate, were measured to determine limiting climatic values when workers need to rest. In a follow-up study, it was recommended that the whole safety management system be re-engineered rather than focusing on piecemeal, incremental improvements (Rowlinson et al., 2014). The practical outcomes of this study led to further comparative studies into institutional logics in different cultures (Jia et al., 2017).

Overall, the vast range of construction studies conducted over the years have been well-meaning and have been implemented with enthusiasm and a desire to improve workers' lots. However, many studies, excepting heat stress studies, have little grounding in terms of theory or science. In general, CMR lacks a sound overarching theoretical perspective within which the piecemeal approaches to identified problems can coalesce into a body of knowledge that is both comprehensive and understandable. Although well-meaning, CMR is often fundamentally flawed, and the results are difficult to implement in practice. From the outside, one might say that the research community in construction OHSW is fragmented and lacks an agenda. Appropriate and different theoretical perspectives need to be explored and a whole range of research methods employed.

Institutional logics

Recent research has taken an institutional logics perspective and investigated the mechanisms by which paradoxical and conflicting objectives, PrP particularly, are manipulated by factions or individuals in construction project teams. Deviance from what is planned or specified can be constructive – improving buildability, for instance – or destructive, substituting inferior goods or proposing inferior alternative designs (e.g. Wang, 2021; Siva, 2022). The construction industry and its bidding systems, as currently in practice, encourage such negative practices, and its general culture is set up for conflict rather than collaboration.

Cornelissen et al. (2020) adopted the concept of local rationality to explain how decision-making at work is based on knowledge, perspective, and understanding of the situation at the time. They explored how the specific dominant logic in the particular setting affected people's assessment of what is (in)appropriate, (ir)rational, and (un)safe in the safety context. They adopted the institutional logics put forward by Thornton et al. (2013): market, profession, corporation, state, family, community, and religion. In their study, they explored the conflicting outcomes of multiple logics and their impact on the rationale for safety in a railroad construction and maintenance organisation. For instance, within the market logic, priority was given to choice of supplier, quality, and profit; for the construction company, safety was an asset for gaining market share. Within the profession logic, safety was left to the independent judgement, professional ethics, and skills of the engineer and safety manager. The rationale behind the corporation logic was that safety was enforced through rules and procedures in a top-down manner, creating a tension with the profession logic. The rationales for safety displayed through the seven logics were found to be conflicting and the cause of misgivings and tension at different levels in the organisation.

In reality, site managers are counted on to enforce safety rules, leaving workers with little input and engagement, creating conflict, and souring relations between site managers and workers. This creates organisational boundaries and conflict rather than collaboration on safety practices. Jeschke et al. (2021) found that the strategies used to deal with the paradoxical logics were sub-optimal, particularly the 'either–or' approach. Conflict between regulation and professionalism was as strong as conflict between regulation and production. This viewpoint presents a realistic explanation of conflict among parties on construction projects and, more generally, in the construction industry. It opens the door for more detailed studies where the paradoxes between the various logics can be explored. This viewpoint can take CMR to a different level of understanding.

In Hong Kong, Naiduwa-Handi (2021), drawing on Nahapiet and Ghoshal (1998), states that construction organisations are characterised by temporality, diffused authority, interactions, and interdependence. Adopting Blau's (1964) theoretical lens of social exchange, she explains how the three dimensions of social capital (structural, relational, and cognitive) in a construction project influence participants' work engagement, safety compliance, and participation.

Her research used a mixed-methods approach and found that both relational and cognitive dimensions of social capital influence work engagement and safety performance. Contrary to popular belief, structural social capital had no impact on work engagement or safety compliance. Structural social capital in a project organisation was perceived by the workers as the responsibility of the parent organisation towards them. Furthermore, 'work-related help and support' was not considered an extended benefit, but rather an organisational responsibility and a set of control mechanisms put upon the workforce to meet production targets. Structural social capital does not directly influence engagement or the in-role performance of individuals. This research is further evidence of the paradoxes of production and safety demands, and adds to our understanding by providing insight into how the workforce perceives management.

It is clear the construction industry is beset by conflicts and paradoxes that trade off OHSW against progress and profit. Whether from an institutional logics viewpoint or a social exchange viewpoint, the outcome is the same: there are unresolved conflicts at many levels within the corporate and project organisation. This conflict needs to be understood at all levels within the industry; second, participants – including clients; the professions; the organisations that design, construct, and maintain our facilities; and the workforce – must accept the need to put OHSW first. Attitude change throughout the organisation, accepting worker engagement, and buying into worker ideas and procedures is central to the change process. Institutional complexity caused by competing institutional logics is continually fed by the social and physical distance between directors and project managers and between project managers and frontline workers. However, we issue a note of caution here to CMR researchers. This approach to institutional logics is critiqued by Johansen and Waldorff (2017). Construction researchers are advised to take on board researchers from the social sciences if they wish to follow such ambitious routes outside of their specific domain of knowledge.

A way forward

Reviewing OHSW research, starting with Heinrich and the manufacturing industries, and moving to Dekker and Reason and industries such as aviation and nuclear power, and to high-risk industries that provide a social service at profit, gives us a foundational understanding of how thinking around safety management has developed. However, the construction industry has adopted an approach that picks and chooses from a range of practices and theories, and pulls these together to develop another partial solution to the industry's problems.

Initially, much of the research into OHSW management was empirical. This was necessary to explore the breadth and depth of the problem and to build up a knowledge base of how and why accidents happen. This analysis of statistics and accident data provided a snapshot of the problem. As researchers delved deeper, they took on board from psychology and sociology how people and groups built their own versions of reality. Recognising that OHSW management focused on not only people and social exchange but also the specific industry technology, there was a move towards more rational thinking, and researchers adopted a socio-technical viewpoint. At this point, research became more focused and probed into the decision-making and rationale behind the processes and systems at work, understanding business as a socio-technical system. This stimulated the viewpoint that workers are not to blame for accidents, but the infrastructure and systems with which they worked were not a good fit in terms of OHSW management. Having moved the debate into this interaction between systems the underlying issues driving most sectors is the balance between delivery of a product on time and at a profit, and the social contract with the workforce. The construction industry, however, is not good at striking this balance.

Recent construction industry research into institutional logics has illuminated issues that the industry has struggled with for decades. The production logic, driven by the client, focuses on delivering a project that is a business and profit provider but is at odds with both the professional code of conduct and the governance structures of OHSW. The construction industry faces a problem in terms of both research and industry development. Until the industry can address the conflicting logics of productivity, professionalism, and governance, it will address the day-to-day issues of achieving safety leading and lagging indicators. The development of a strong group safety climate and company safety culture is essential to develop a safety management system that is fit for purpose.

The unresolved issue of incompatible institutional logics makes it almost impossible to pull together a safety management system that is credible and viable. Hence, the way forward would seem to be to direct research into the mechanisms and underlying principles of the client's project dilemma. That dilemma, pointed out at the beginning of this chapter, is that the client wishes to have a facility that is profitable (in whatever way that is defined), while the construction industry, from designers through to facilities managers, compete as a dysfunctional team to win projects at, in most instances, the lowest price.

In short, PrP delivers a value chain that will not deliver OHSW to the workforce. Hence, a sea change in the way the industry operates is necessary at the industry level, the stakeholder level, the project level, and the individual work packages on the construction site. If progress is to be made, the management and business models of construction firms need to be addressed by CMR. Research needs to be refocused on what has been done in the wider OHSW discipline so that the well-meaning research has greater merit, impact, and validity. It is laudable that those who research in this field have a passion for improvements in the industry and wish to make both workers and project users safe. However, by choosing small areas of incremental improvements without addressing the overarching institutional issues, research impact will be piecemeal, narrowly focused, and academically weak compared with other disciplines.

References

Albert, A., Hallowell, M.R., Kleiner, B., Chen, A. and Golparvar-Fard, M., 2014. Enhancing construction hazard recognition with high-fidelity augmented virtuality. *Journal of Construction Engineering and Management*, 140(7), 04014024.

Amalberti, R., 2001. The paradoxes of almost totally safe transportation systems. *Safety Science*, 37, 109–26.

Banwell, Sir Harold, 1964. *The placing and management of contracts for building and civil engineering work*. London, HMSO.

Bhandari, S., Albert, A. and Hallowell, M.R., 2020, November. Construction hazard recognition: Themes in scientific research. In *Construction Research Congress 2020: Safety, workforce, and education* (pp. 58–66). Reston, VA, American Society of Civil Engineers.

Bird, F. E. Jr., 1966. *Damage control: A new horizon in accident prevention and cost improvement*. New York, American Management Association.

Blau, P.M., 1964. *Exchange and power in social life*. New York, Wiley.

Cherns, A.B. and Bryant, D., 1984. Studying the client's role in construction. *Construction Management and Economics*, 2(2), 177–84.

Cooper, M.D., Phillips, R.A., Sutherland, V.J. and Makin, P.J., 1994. Reducing accidents using goal setting and feedback: A field study. *Journal of Occupational and Organizational Psychology*, 67(3), 219–40.

Cornelissen, P.A., Van Vuuren, M. and Van Hoof, J.J., 2020. How logical is safety? An institutional logics perspective on safety at work. *Work*, 66(1), 135–47.

Dekker, S.W., 2003. Accidents are normal and human error does not exist: A new look at the creation of occupational safety. *International Journal of Occupational Safety and Ergonomics*, 9(2), 211–18.

Gibb, A., Haslam, R., Gyi, D., Hide, S. and Duff, R., 2006, November. What causes accidents? *Proceedings of The Institution of Civil Engineers-Civil Engineering*, 159(6), 46–50.

Haddon Jr, W., 1973. Energy damage and the ten countermeasure strategies. *Human Factors*, 15(4), 355–66.
Hallowell, M.R., Hinze, J.W., Baud, K.C. and Wehle, A., 2013. Proactive construction safety control: Measuring, monitoring, and responding to safety leading indicators. *Journal of Construction Engineering and Management*, 139(10), 04013010.
Harvey, E.J., Waterson, P. and Dainty, A.R., 2018. Beyond ConCA: Rethinking causality and construction accidents. *Applied Ergonomics*, 73, 108–21.
Heinrich, H.W., 1931. *Industrial accident prevention. A scientific approach.* New York, McGraw-Hill.
Hillebrandt, P.M., 2000. *Economic theory and the construction industry.* London, Macmillan.
Hillebrandt, P.M. and Cannon, J., eds., 1989. *The management of construction firms: Aspects of theory.* London, Macmillan.
Hinze, J., 1997. The distractions theory of accident causation. *CIB REPORT*, 112–21.
Hollnagel, E., 2006. Accident analysis and human error. In Informa Healthcare & W. Karwowski (eds.), *International Encyclopedia of Ergonomics and Human Factors*, pp. 1889–92. New York, Routledge.
Hopkins, A., 2007. *The problem of defining high-reliability organisations.* Vancouver, National Research Center for Occupational Safety and Health Regulation.
Hopkins, A., 2021. *A practical guide to becoming a 'high-reliability organisation'.* Canberra, Australian Institute of Health & Safety.
Jeschke, K.N., Waldorff, S.B., Dyreborg, J., Kines, P. and Ajslev, J.Z., 2021. Complaining about occupational safety and health: A barrier for collaboration between managers and workers on construction sites. *Construction Management and Economics*, 39(6), 459–74.
Jia, A.Y., Rowlinson, S., Loosemore, M., Xu, M., Li, B. & Gibb, A., 2017. Institutions and institutional logics in construction safety management: The case of climatic heat stress. *Construction Management and Economics*, 35(6), 338–67.
Johansen, C.B. and Waldorff, S.B., 2017. What are institutional logics and where is the perspective taking us? In G. Krücken, C. Mazza, R.E. Meyer, & P. Walgenbach (eds.), *New themes in institutional analysis*, pp. 51–76. Northampton, MA, and Cheltenham, UK, Edward Elgar.
Lingard, H. and Rowlinson, S., 1998. Behaviour-based safety management in Hong Kong's construction industry: The results of a field study. *Construction Management and Economics*, 16(4), 481–8.
Lingard, H., Pirzadeh, P. and Oswald, D., 2019a. Talking safety: Health and safety communication and safety climate in subcontracted construction workgroups. *Journal of Construction Engineering and Management*, 145(5), 04019029.
Lingard, H., Oswald, D. and Le, T., 2019b. Embedding occupational health and safety in the procurement and management of infrastructure projects: Institutional logics at play in the context of new public management. *Construction Management and Economics*, 37(10), 567–83.
Nahapiet, J. & Ghoshal, S. 1998. Social capital, intellectual capital, and the organizational advantage. Academy of *Management Review*, 23, 242–66.
Naiduwa-Handi, C. 2021. *Effects of project social capital on safety performance: The mediating role of work engagement*, PhD thesis. The University of Hong Kong, Hong Kong.
Oswald, D., Ahiaga-Dagbui, D.D., Sherratt, F. and Smith, S.D., 2020. An industry structured for unsafety? An exploration of the cost-safety conundrum in construction project delivery. Safety *Science*, 122, 104535.

Oswald, D., Borg, J. and Sherratt, F., 2019. Mental health in the construction industry: A rapid review. In C. Pasquire and F.R. Hamzeh (eds.), *Proceedings of the 27th Annual Conference of the International Group for Lean Construction*, pp. 1049–58, Dublin, Ireland, International Group for Lean Construction.

Oswald, D. and Lingard, H., 2019. Development of a frontline H&S leadership maturity model in the construction industry. Safety *Science*, 118, 674–86.

Oswald, D., Sherratt, F. and Smith, S., 2017, September. An investigation into a health and safety rewards system on a large construction project. In P.W. Chan & C.J. Neilson (eds.), Proceedings of the 33rd annual ARCOM Conference, pp. 370–9. Cambridge, UK, ARCOM.

Oswald, D., Sherratt, F., Smith, S. and Dainty, A., 2018. An exploration into the implications of the 'compensation culture' on construction safety. *Safety Science*, 109, 294–302.

Patching, A., 2022. *Attitudes to psychological stress among construction professionals.* London, Springer Nature.

Petersen, D., 2003, December. Human error: A closer look at safety's next frontier. Professional Safety, 48, 25–32.

Reason, J. 2000a. Human error: Models and management. *British Management Journal*, 320(7237), 768–70.

Reason, J., 2000b. Safety paradoxes and safety culture. Injury Control and Safety Promotion, 7(1), 3–14.

Reason, J. and Hobbs, A., 2003. Managing maintenance error: A practical guide. Boca Raton, FL, CRC Press.

Robertson, L.S., 1992. *Injury epidemiology*. Oxford University Press.

Rowlinson, S. & Jia, Y.Y.A., 2013. Application of the Predicted Heat Strain (PHS) model in development of localised thresholds-based heat stress management guidelines for the construction industry. Annals of Occupational Hygiene, 58(3), 326–39.

Rowlinson, S., Jia, Y.Y.A., Li, B. and Ju, C., 2014. Management of climatic heat stress risk in construction: A review of practices, methodologies, and future research. *Accident Analysis & Prevention*, 66, 187–98.

Simon, H., 1969. *The sciences of the artificial*. Cambridge, MA, MIT Press.

Siva, J. P. S., 2022. *An examination of power relations in megaproject decision-making: application of governmentality theory*, PhD Thesis. Newcastle University, Australia.

Sulzer-Azaroff, B., 1980. Behavioral ecology and accident prevention. *Journal of Organizational Behavior Management*, 2(1), 11–44.

Thornton P.H., Ocasio W. and Lounsbury M., 2013. *The institutional logics perspective: A new approach to culture, structure, and process*. Oxford, Oxford University Press.

Tixier, A.J.P., Hallowell, M.R., Rajagopalan, B. and Bowman, D., 2016a. Automated content analysis for construction safety: A natural language processing system to extract precursors and outcomes from unstructured injury reports. *Automation in Construction*, 62, 45–56.

Tixier, A.J.P., Hallowell, M.R., Rajagopalan, B. and Bowman, D., 2016b. Application of machine learning to construction injury prediction. *Automation in Construction*, 69, 102–14.

Trist, E. and Bamforth, W., 1951. Some social and psychological consequences of the long wall method of coal-getting. *Human Relations*, 4, 3–38.

Walker, G.W., 2010. A safety counterculture challenge to a 'safety climate'. *Safety Science*, 48(3), 333–41.

Wang, H., 2021. *Transforming practices over large institutional distances: The case of a Chinese construction firm*, Ph D thesis. Reading, UK, Reading University.

Weick, K. & Sutcliffe, K., 2001. *Managing the unexpected: Assuring high performance in an age of complexity*. San Francisco, Jossey-Bass.
Weick, K., Sutcliffe, K. and Obstfeld, D., 1999. Organising for high reliability: Processes of collective mindfulness. *Research in Organisational Behaviour*, 21, 81–123.
Woods, D.D. & Cook, R.I., 2002. Nine steps to move forward from error. *Cognition, Technology & Work*, 4, 137–44.
Woodward, J., 1965. *Industrial organization: Theory and practice*. Oxford, Oxford University Press.
Zohar, D., 1980. Safety climate in industrial organizations: Theoretical and applied implications. Journal of *Applied Psychology*, 65(1), 96.
Zohar, D., 2000. A group-level model of safety climate: Testing the effect of group climate on microaccidents in manufacturing jobs. Journal of *Applied Psych*ology, 85(4), 587.

9 A research agenda for construction management in the 4.0 era

Evangelos Pantazis, Eyüphan Koç and Lucio Soibelman

Introduction

The construction industry is undergoing a transformation in the way work is performed, delivered, and managed. This is largely due to the level of maturity that various technologies have reached, which in turn accelerated the cultural change in the construction industry. Owners are demanding higher performance buildings delivered at lower costs and in shorter time frames. Policies are being devised for meeting the United Nations Sustainable Development Goals and reducing the energy footprint of the built environment. Meanwhile, digital innovation is diffusing more than ever into the industry.

Digitization and automation in AEC (Architecture, Engineering, Construction) have been happening at different rates across disciplines; however, it has been evident that integrated innovation is required to leverage the benefits of these advancements more rapidly and at a larger scale. The benefits of integrated innovation have become evident not only (1) via the observation of other industries, such as manufacturing, but also (2) via innovative competition coming from the new actors (i.e., tech industry) in the AEC industry.

Today, as AEC stakeholders experience a stronger push for innovation in an effort to capitalize on the digital know-how of each discipline synergistically, improving key project performance indicators on time, cost, productivity, safety, and so on remains the main target. Meanwhile, although standardization and reductionism fulfilled their promise to deliver more affordable buildings globally, they failed in reducing the waste in construction processes and in supporting environmental sustainability. To excel in the digital age, the construction industry specifically and the AEC as a whole need to reexamine their tools, methods, and processes closely to search for new ways that help reduce waste and increase productivity, which have been stagnant for the last 30 years.

Work completed in the last decades in academia and in practice has shown the promise of feeding designers with information on the physical implementation of their design during the early design stages (Tzortzopoulos et al., 2020). It is also accepted that there is a need to co-design the *product* (building) with the *process* (how to use it, what is to be built, how to build it) and to integrate value engineering tools in the design process (Sriram et al., 2015; Tzortzopoulos et al., 2020). The lack of a holistic vision on how AEC should transform itself deemed the surge in the development of tools and technologies ineffective in terms of questioning the existing construction paradigm. This meant that most new technology focused on the automation of tasks within the existing production paradigm.

Originally developed to address the mentioned lack of a holistic vision, the principles of lean construction, introduced as early as the 1990s, were not adopted at a large scale in the industry for the following reasons. They failed to (1) address the fragmentation challenges involved, and (2) handle the complexity of the socio-economic parameters affecting the adoption of innovations. However, a recent wave of disruptions originating outside the AEC has emphasized an opportunity to address the challenges, while successful applications of lean methods in a number of cases have indicated their potential in facilitating the shift from fragmentation to integration in practice. This shift can be further enabled by a more widespread adoption of lean principles in AEC, which can help utilize technological advancements recently grouped under the framework "Construction 4.0" holistically (Sawhney et al., 2020a).

Understanding the changing socio-technical needs of AEC practice and emerging requirements for the workforce will be a critical determinant of success in the mentioned shift. We touch on these aspects in the next section, which is followed by a brief overview of key concepts and technologies in Construction 4.0. We describe the implications of Construction 4.0 technologies for the AEC sector and particularly in relation to the field of on-site construction management (CM). Later in the chapter, a research agenda for CM is laid out to discuss how to address existing challenges, touching on multiple research problems in data-driven and automated project management (e.g., scheduling, cost estimating), digital modeling (e.g., Scan-to-Building Information Modeling/Management [BIM], digital/automated inventory and supply chain management, integrated project delivery [IPD] as a common-pool resource, sustainability and industrialized construction, management of digital fabrication, etc.). The chapter ends with a discussion on how production needs to be rethought as flows of information and not merely as a transformation of inputs and outputs, and how lean thinking has the capacity to reformulate research and educational agendas to better tackle longstanding problems in the AEC.

Changing socio-technical needs of AEC practice and emerging requirements for the construction workforce

Given the backdrop above, the transformation in the roles of architects and engineers due to the ongoing 4.0 revolution is a point of emphasis particularly motivated by the disruption created by actors coming into AEC from other industries. In the last decade, the mentioned disruption was led by tech firms that promised to reinvent construction by vertically integrating the heavily fragmented supply chain and revolutionizing project delivery.

The push by the newcomers (e.g., Katerra, Sidewalk Labs) showed with varying levels of success how important it is to rethink the AEC holistically and to reconsider the set of tools and skills required for all actors within the industry (architects, engineers, construction managers, contractors, etc.), but also to redefine their relationships. If the CM workforce is taken as an example, they predominantly carry out the traditional tasks of scheduling, estimating, controls, and so on, where individuals develop expertise within one such task and gather experience, allowing them to charge more for their labor. At a higher level, project managers currently need to gather information from multiple sources (i.e., emails, reports, etc.) and merge it into a bigger picture. In parallel, they need to manually update plans and communicate new versions to team members and report the progress to a superior/executive. It is established that the mentioned processes reduce productivity on all levels in the organization and make it difficult to have a holistic overview of the project when the information is fragmented.

Observing such longstanding challenges and the wave of digitization/automation already altering individual tasks in CM, the question immediately becomes: Will construction managers be doing cost estimation and scheduling in 10 or 20 years, or will they be replaced by data and Artificial Intelligence (AI)-driven tools that can process extensive amounts of current (productivity, weather, supply chain, etc.) and historical data (i.e., digitized human experience)? In other words, why should we consider the CM workforce or AEC workforce in general to be *robot (automation) proof*, in Aoun's (2017) words, while other industries have persistently seen continuous improvement in their production processes by integrating computation and automation?

What becomes the job of the construction manager when a fleet of Unmanned Aerial Vehicles (UAVs) and other on-the-ground robots can autonomously navigate a construction site and accurately measure and report productivity? When inventories are digitized and automated from procurement to installation? And when schedule breakdowns or cost overruns are predicted accurately and continuously? Perhaps most significantly, what is the job of the construction manager when all such tasks are integrated and coordinated digitally through data-intensive, accurately representative, behaviorally rich digital models?

Naturally, there will always be obstacles in front of such advancements, but the big push over the last decade hints that a large-scale modernization of the AEC industry as a whole is not far off. We assert that the key is to understand the potential impacts of the new modes of production promoted within the 4.0 framework on the industry and its workforce to collectively adapt to the change. Today, are construction professionals prepared to lead the development of novel "integrated" tools and methods that are likely to be software and data intensive? Or are they rather unsophisticated followers of whichever new tool or technology is put forward by organizations led by software engineers? Today, academic programs in construction do not teach the skills to build these systems and processes. This exposes a substantial portion of industry professionals, from young graduates to senior managers with 30–40 years of experience, to a situation where people with more diverse training and experience could come in and replace them.

We do not argue that all tasks and jobs should be protected at any cost when they are challenged by digital transformation. On the contrary, the argument is that such a transformation could and should be led by the AEC workforce, which requires that there is an ongoing discussion considering how the workforce will be trained in university programs. A good example is BIM, which is deployed as an all-inclusive automated quantity take-off tool today while still being behind on cost estimating, particularly because it does not accommodate the knowledge and experience accumulated by experienced cost estimators. Such knowledge and experience are often specific to location, project type, the contractor, and so on, so is it reasonable to expect that there will be an "update" to the software to carry out cost estimating tasks as accurately as humans around the globe at all times? It is clear that experienced CM professionals must lead the way in automating cost estimation for their firms, ideally by customizing software or by programming modular pieces of software that can be integrated into other software (e.g., BIM).

In this section, our initial objectives are to (1) provide a snapshot of Construction 4.0 concepts and technologies, and (2) briefly analyze the implications of these technologies in terms of shaping the future of the industry. Providing a broad presentation of this area will form the foundation of the discussion in the following sections. Despite references to specific technologies and advancements, the emphasis is on the overarching insights.

"Construction 4.0" can be broadly considered as a "transformative framework" that encompasses technologies and concepts as they relate to the AEC industry (Sawhney et al., 2020b). The exponential increase of publications on this topic in the last five years can testify to the evolution of and interest in the term, the expanding spectrum of concepts that it encompasses, as well as to the speed of the transformation that the AEC industry is undergoing. The concepts and technologies can be divided into three main categories: (1) smart construction site (smart factory), (2) simulation and modeling, and (3) digitization and virtualization (Oesterreich & Teuteberg, 2016). One taxonomy of these concepts and technologies is given in Figure 9.1, which additionally demonstrates the level of innovation diffusion (i.e., five levels: awareness, interest, evaluation, trial, adoption) corresponding to each concept or technology today and as expected in a future of more integrated innovation and project delivery.

Construction 4.0: key concepts and technologies, and implications for the industry

Key concepts and technologies

For example, BIM, Mobile Interfaces, Internet of Things and Services (IoT/IoS), and Cloud Computing have reached the level of adoption even within traditional construction practice, while technologies such as High-Performance Computing and Autonomous Robotics, and concepts such as Project Life Management are still regarded to be at the level of awareness. In between, there is a group of concepts that includes Energy and Construction Simulations, Large-scale Additive Manufacturing, Data-Driven Generative Design Tools, Embedded Sensors/Radio Frequency Identification (RFID), Virtual/Augmented Reality (VR/AR), and UAV, and techniques including big data analytics, modularization/prefabrication, and robotics and automation. As mentioned, these concepts and technologies can be considered within three broad categories based on their respective field of application: modeling and simulation, digitization and virtualization, and smart construction site (both on site and off site; Sawhney et al., 2020b). It is essential to note that Figure 9.1 represents projections that are based on our expectations with regard to the future of the industry and the outlook of 4.0-enabled project delivery. Arriving at that future will require a substantial amount of work on the industry and

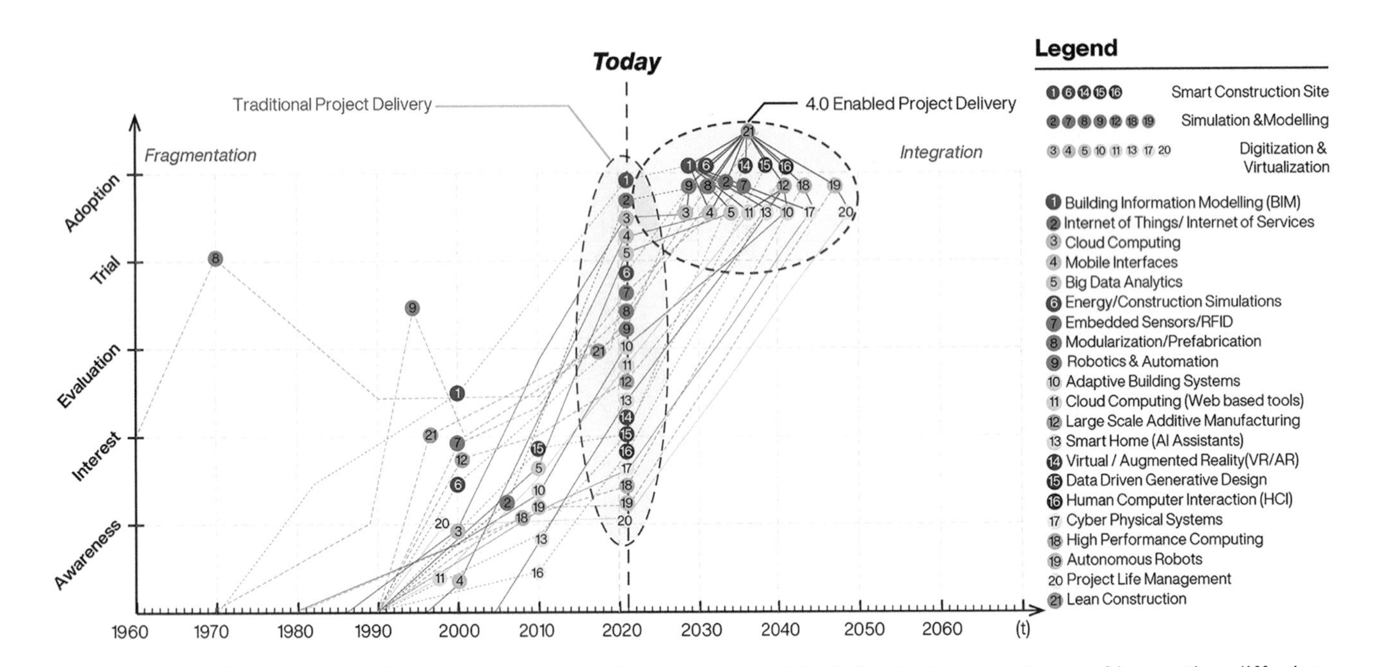

Figure 9.1 Construction 4.0 concepts and technologies in terms of their "paths" across stages of innovation diffusion

academic fronts. In the next subsection, we discuss factors deeming this transformation more difficult.

Construction management in transition: challenges in the 4.0 era

As early as the beginning of the 1990s, traditional construction planning and organization was challenged by lean construction, which quickly distinguished itself both in terms of the principles it puts forward, and the tools and methods for its implementation. Circumstances that are specific to AEC, such as the fragmented nature of the industry, combined with a strongly risk-averse culture that is slow to adopt innovation were forces that did not allow a rapid adoption of new methods and tools that lean construction research put forward (Forcael et al., 2020). Today, capital projects are still largely managed according to the traditional understanding of the cost, time, and quality tradeoff. Fragmentation is still an obstacle in front of (digital) innovation (Koç et al., 2020), and the industry could still be regarded as risk-averse in terms of adopting new concepts, tools, and technologies. Issues of low productivity and waste persist and are perhaps under more scrutiny due to rising sustainability and resilience standards concerning the built environment. There is an increasing number of start-up firms attempting to tackle the chronic issues of the industry through the use of cutting-edge technologies. The quick rise and fall of Katerra, which focused on revolutionizing construction through a full vertical integration, shows that there is no clear understanding of the local vs. global inefficiencies of the AEC industry. AEC being fragmented (horizontally, vertically, longitudinally) is typically perceived as a negative, but it also means that parts of the industry (and its supply chain) are running very efficiently. Katerra's goal to try to radically change everything in the construction industry rather than focusing on restructuring the domains that needed repair proved to be catastrophic. The company was unable to generate the projected efficiency through vertical integration and was eventually forced to file for bankruptcy despite attracting large sums of investments. One striking aspect is the economies of scale benefits that Katerra expected to achieve, but due to the inconsistent demand for construction, they could not engineer a supply chain that allowed them to capture the mentioned benefits, which in turn affected their prices adversely. It becomes evident from the Katerra example that there is no way forward for a venture in revolutionizing the industry before the fundamentals are right (i.e., understanding the client and the market). Further, in the case of Katerra, according to its staff, the failure was not due to attempting to apply innovative techniques and offsite fabrication, but to unchecked growth, ill-positioned staff, and undefined goals (Davis, 2021). The Katerra story shows once more that the transition from theory to the implementation of 4.0 ideals in construction is complicated, and there is still ground to be covered, particularly on the theoretical approaches relating to technology-enabled production management. At this junction, the following questions arise: Could the teachings of lean construction be a catalyst for digital innovation in the AEC? Is there potential for a symbiosis where lean principles and concepts pave the way for integral 4.0 innovations, concepts, and technologies that enable a more effective implementation of lean principles and concepts?

Learning from the past

Lean construction concepts were developed almost concurrently with their counterparts in manufacturing, but due to the slow adoption of technology in the AEC as well as the premature status of the technologies themselves, the lean approach was not able to address real project complexity and, thus, did not "make it" in the industry (Tommelein & Ballard, 1999). Only a handful of concepts have been successfully implemented into methods such as the Last Planner System (LPS). LPS has been proven to be an efficient method of project management and control but has not been widely adopted in the industry. On the other hand, the traditional activity-centered view of construction demanded the use of methods such as the Critical Path Method (CPM) for scheduling and project controls. However, waste activities, such as waiting, storing inventory, moving material, and inspection, are not generally considered by traditional tools and methods. Additionally, the current construction paradigm assumes that the productivity of the whole project can be improved by incrementally improving the productivity of each activity. On the contrary, in manufacturing, changing the production paradigm resulted in moving away from the activity-centered paradigm and motivated a focused effort on improving competitiveness by identifying and eliminating waste (non-value-adding) activities.

Recent research in the field of Construction 4.0 has mainly focused on technological areas, such as big data, BIM, and AI (Forcael et al., 2020), and less on theoretical approaches that can help encompass all these innovations. Construction 4.0 tools that are used today are project and object based. BIM, largely regarded as the centerpiece of design and CM in a 4.0 future, organizes information around construction objects (e.g., components, products) and not around the set of processes contributing to the final construction of those objects (e.g., production, transportation, on-site storage, assembly). BIM data organized around objects do not currently allow the deployment of AI workflows that would permit one to correlate processes within and across projects. Such workflows could enable the holistic modeling and simulation of operations for construction companies. The missing piece is common data structures (i.e., ontologies) to represent all processes and products that would allow the fusion of data across objects, processes, and projects into data warehouses in which advanced analytical tools can be developed. Lean construction, in contrast with activity-centered, traditional CM, emphasizes a process-based view (i.e., a focus on flows of information and material), and allows the generation and management of process-based data through the deployment of the tools it recommends (e.g., LPS, KanBIM). Thus, it can support the development of the missing data structures mentioned above, paving the path to enhanced analytics and widespread use of AI.

Implementing lean construction in this data-driven manner requires the use of emerging technologies and concepts related to the Construction 4.0 framework. Therefore, this two-way relationship will be central to moving the industry beyond the obstacles that hindered the modernization of AEC in the past.

Lean thinking in construction

Broadly speaking, lean thinking has as a primary goal to better meet customer needs while using less of everything (So & Sun, 2010). Its main objectives include the design and implementation of production systems that can deliver custom products instantly on order, but that maintain no intermediate inventories. In manufacturing, where typical standardization approaches have focused on keeping the machines running and the production line moving to minimize the cost of each part, the Toyota Production System (TPS) focused on setting a multi-dimensional standard of perfection that prevents sub-optimization and allowing continuous process improvements to be made instead of repeating an established process (Liker & Meier, 2006). Lean thinking, which emerged in the 1990s, is manifested with TPS and calls for a paradigm shift from "batch and queue" mass production systems to product-aligned "single-piece pull-based" production systems (Table 9.1). The idea behind single-piece pull-based manufacturing is to eliminate the inventory and, hence, the safety net, and expose the system inefficiencies to build a robust, supple, efficient manufacturing system ready for the challenges posed by the globalization of the markets (Womack & Jones, 1997).

Despite its presence for more than 30 years, there seems to be no clear theory of lean. This is because Japanese culture, where it originated, tends to emphasize direct personal experience (to learn something) rather than abstract theories (Tzortzopoulos et al., 2020). Thus, the TPS does not have a clear theoretical base, but it emerged as the unplanned and unexpected result of seemingly unrelated innovations, improvements, and initiatives that resulted in a set

Table 9.1 Principles, philosophies, and methods of traditional vs. lean construction paradigms

Paradigm	Traditional Construction	Lean Construction 4.0
Principles	• Activity-oriented project control	• Value-oriented project control
Philosophy	• Aim for "good enough" (i.e., to stand)	• Aim for perfection (continuous improvement)
Organizational philosophy	• Vertical: each discipline takes responsibility within the bounds of their contract	• Horizontal: value streams using appropriate level of empowerment, pushing responsibility across all phases of the project
Methods	• Top-down: managers decide the planning of the processes on site	• Bottom-up: people working on site decide on planning of processes on site
People - AEC	• Segregation of architects, engineers, and builders in every phase	• Teams of multi-skilled architects, engineers and builders at all levels of a project
People - Construction Management	• Scheduling • Costing (add up and manage costs) • Estimating • Project control (CPM)	• Project-delivery strategy • Reduce waste in processes • Predictive costing • Orchestrating different actors

of best practices. Another reason for the lack of a clear theory of lean is that management research has been decisively influenced by the conception that production was – just – an application context for managerial methods rather than an object of research. What this means is that production was not considered within the domain of management (Gordon & Howell, 1959; Koskela et al., 2017). In the context of lean, production becomes a process of continuous improvement by tightly co-relating flows of information, people, and available resources.

All embodiments of lean, in car manufacturing, healthcare, and other fields, have been arguably contextual in the sense that the characteristics and specific problems of each industry are mirrored in the lean production model, methods, and tools. Lean construction could be defined as the contextual application of generic principles of lean thinking, mainly originating from car manufacturing

(Gao & Low, 2014), but further research on the topic has also provided further theoretical grounding, as will be shown in the following section.

At the outset, it is worthwhile to note that specific differences between industrial contexts have been impediments to the wide adoption of lean construction. It is well known that construction has its peculiarities, such as one-of-a-kind/craft production, on-site production in a dynamic environment, and temporary organization, which contrasts with mass production in a factory by a permanent organization (Table 9.1). Thus, lean construction and lean manufacturing are different but comparable: they are desired outcomes of the application of (broadly) similar principles and methods in different contexts. For example, the current form of production management in construction is derived from the same activity-centered approach found in mass production and project management of manufacturing in the 1960s. It assumes that optimization of a project can be achieved by improving the productivity of each activity, assuming customer value has been identified in the design. By breaking down the project into pieces (i.e., design, engineering, and construction) using the work breakdown structure (WBS) method, and then reassembling these pieces in a logical sequence via the CPM, project managers can estimate the time and resources required to complete each activity and, therefore, to complete the project. Each activity is further decomposed until it is contracted to a foreman or a task leader. Although this approach appears to be creating efficiency at the planning level, in many cases it fails to deliver effective results, and the critical path is seldom followed.

A meta-analysis of research into production in construction covering the period from 1970 to 2000 reported that almost 49.6 percent of time was wasted in non-value-adding activities, which are defined by the activity-centered approach mentioned above (Horman & Kenley, 2005). Additionally, a study by Koch (2005) identified that the main factors for failures encountered in construction were production planning and control, communications and cooperation, and design activities. Managing construction under the current paradigm is conceived as monitoring each contract or activity against the schedule and the budget projections. However, managing construction following a lean approach is different because it has a clear set of objectives for the delivery process, and it is aimed at maximizing value for the customer at the project level. Moreover, the product (i.e., building) is designed concurrently with the process and not a priori. Additionally, production control is applied throughout the life of the project instead of being applied only during design development.

To provide a clear overview on lean construction and how it relates to emerging trends and technologies related to the Construction 4.0 framework, an overview of the main principles along with a set of prominent methods will be provided in the following section (Figure 9.2).

The main principles of lean construction: transformation-flow-value

One of the most significant contributions that lean thinking brought to construction was a greater focus on customers. In production, the rationale of the value-generation concept is that the value of a product can be determined only in reference to customers, and the goal of production is to satisfy customers' needs (Levitt, 1960). In this context, generating value means delivering what the customers want, when they want it, and in the amount that they want. Lean construction aims to bring this concept into the built environment where, in a lot of cases, it is assumed that the value is generated by the design of a new building without always verifying that with the customer (i.e., the building occupants). What becomes critical in lean construction is how you define value and for whom, but also how you communicate information related to the delivery of such value. Koskela (1992) attempted to address this in construction via the introduction of a new production philosophy that was manifested in the Transformation, Flow, and Value (TFV) theory.

Koskela's approach extends the theoretical understanding of production to two other theories – the flow theory and the value-generation theory – while it makes use of transformation theory when appropriate (Koskela et al.,

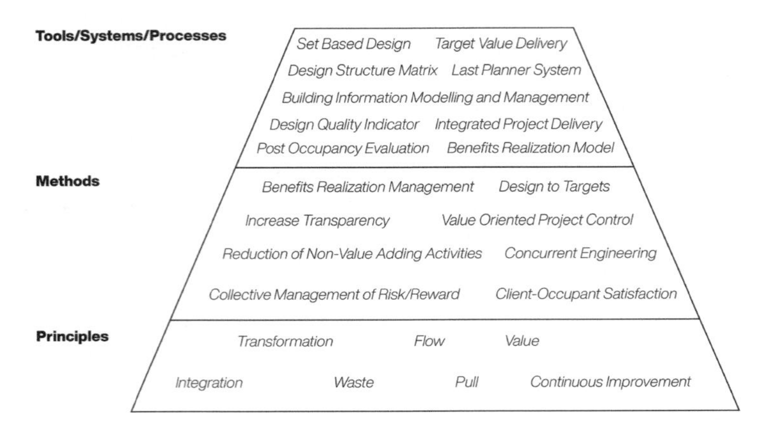

Figure 9.2 The basic building blocks of lean construction, organized from the more abstract (concepts) to the more practical (tools)

2002; Tzortzopoulos et al., 2020). To provide a better understanding of this approach, a brief description of each of the three main pillars is provided.

Transformation theory is central to production management and refers to the conversion of inputs into outputs through an iterative decomposition of tasks. The decomposition entails that the total productive task (i.e., delivering a building) is successively broken down into smaller tasks, until they can be assigned to operatives or companies in the supply chain. This decomposition is based upon two assumptions: a) the tasks emerging from the decomposition are, by their nature, an input–output transformation similar to the original task, and b) the decomposed tasks are mutually dependent. Although these assumptions make management easy, they assume an idealized reality where there are no tasks that depend on other external tasks.

Flow theory deals with understanding design as a flow of information that can lead to reducing waste by minimizing non-value-adding activities, thus seeking the potential to increase efficiency. Non-value-adding activities are considered to be those that are part of a production process but that do not address customers' (i.e., building occupants') needs. By evaluating activities via this perspective, it has been shown that efficiencies can be achieved that include but are not limited to eliminating wait time for information to become available and reducing time for inspecting alternative design solutions, which in turn can reduce design rework (i.e., reviewing the design solutions according to the requirements). Moreover, flow theory can help address the lack of effective planning and control so that the influence of complexity and uncertainty of construction can be reduced by supporting design decision-making via the supply of adequate and consistent information.

Value-generation theory deals with understanding the value delivered to a customer by a product or a service and attempts to identify design criteria and customer needs to be used as inputs to the process. Past research has criticized the predominant understanding of value generation in the construction industry (Winter & Szczepanek, 2008). Winter and Szczepanek (2008) argue that project management practices are traditionally focused on the creation and delivery of physical products to specification, cost, and time, without much consideration of how these physical assets will ultimately contribute to fulfill a purpose or intent. In the context of construction, we need to consider that the nature of perceived value is complex, multi-dimensional, and implies an interaction between a customer/user and other stakeholders (e.g., city council, building code) and a product (i.e., a building project). Additionally, value is relative by virtue of its comparative, personal, and situational nature. Lastly, value is preferential, perceptual, and cognitive–affective by nature. In many cases, value is associated with value engineering, which is tightly connected to construction cost reduction. However, creating value should be more important than minimizing design costs, as the impact of any investment in design is likely to be greatly outweighed by delivering a building that supports its users' needs and, hence, supports business profitability.

The innovation of lean construction lies not in the recontextualization of methods and tools from the TPS in the field of construction, but rather in theoretically extending the traditional concept of production with the TFV framework. Lean construction emphasizes the importance of design as the main means to generate value to end-users by considering it not simply as a process that transforms materials into physical objects, but rather as a process in which information is transformed into physical objects while the value added is continually informed by clients. Lean thinking highlights the importance of integrating design and production, from the conceptual phase throughout the whole development and operation stage, but also emphasizes the need to manage design and production holistically.

What is important to note is that this integration is promoted not by technological advancements but as a principle that can ensure the delivery of more value to the customers and the built environment as a whole (e.g., by delivering buildings that are not harmful to the environment). In the traditional CM paradigm, the knowledge of delivering value appears to be possessed by the domain experts (i.e., architects, engineers, and project managers) who are thus able to provide executable production plans. On the contrary, in the lean paradigm, the knowledge of delivering value is distributed among clients, practitioners, and domain experts who are working in unison to create continuously improving production processes. In the following section, we will discuss tools and methods that illustrate how the concepts can be applied in practice.

Methods for lean project management: a toolkit for lean construction in the 4.0 era

Although the theoretical basis of lean is obscure, the tools and methods are well established. In Table 9.2 we describe important tools and how they could be related to lean construction in the 4.0 era. Some of these tools might not be new to readers, as they have existed for some years and have been extensively described in the lean construction literature (Koskela et al., 2002; Tzortzopoulos et al., 2020). Nonetheless, it is important to reevaluate them in light of recent technological advancements.

Table 9.2 A collection of lean construction techniques that could be utilized in conjunction with Construction 4.0 concepts and technologies

Tool	Description	Potential Use
Benefits Realization Model (BeReal)	A technique that suggests continuous cycles to support value generation: a) envisioning outcomes, b) implementing action to achieve outputs, c) checking intermediate results, and d) dynamically adjusting the path leading from investment to outcomes	When developing the project charter and identifying the stakeholders
Building information Modeling/ Management (BIM)	A term that encapsulates a number of benefits from the adoption of BIM tools (e.g., reduced design development life cycle; effective capture and flow-down of intent; reduced rework; increased iterations for value improvement)	Throughout the project's life cycle, from initiation to delivery and operation
Choose by Advantage (CBA)	A decision-making system that evaluates the impacts of different advantages and uses them to drive design decisions	At defining moments of the project (e.g., building permit, initiation of construction)
Concurrent Engineering (CE)	The concept of simultaneously considering multiple requirements from diverse stakeholders, which leads to performing activities that may be overlapping	When proceeding from design development to construction documents
Design Structure Matrix (DSM)	A method for information flow representation where tasks are organized in rows and columns according to their expected chronological order	A data collection exercise that can take place when mapping requirements to tasks and activities
Design Quality Indicator (DQI)	A method that assists stakeholders in defining priorities and assessing the quality of design based on the three fundamental Vitruvian Principles: durability, utility, and aesthetics	When proceeding from schematic design to design development
Integrated Project Delivery (IPD)	A project-delivery approach that integrates people, systems, business structures, and practices and aims to reduce waste and optimize efficiency through all project phases	When the contractual relationships of a new project are defined

Tool	Description	Potential Use
Last Planner System (LPS)	A method that models the instability of the various flows at the level of detail necessary for effective production controls	Throughout the execution of the construction
Set-Based Design (SBD)	A method for generating and considering sets of alternative solutions from the beginning of the design process instead of fixating on one alternative and developing it in detail	When developing schematic designs at the early design stage
Target Value Delivery (TVD)	The practice of steering design to targets using costs as a driver of design and not because of it, constraining the design and building a facility to a maximum cost	Helps project managers to achieve cost predictability by focusing on market prices and accomplishing returns

The tools and methods of lean construction span all phases of a project, from conceptual design – such as Set-Based Design (SBD) and Target Value Delivery (TVD) – to construction – such as IPD – to operations – such as Post Occupancy Evaluation. Their main objective is to create a more efficient design and CM process. The potential uses listed in Table 9.2 describe ways to help manage and control projects by establishing and pursuing a clear set of objectives, with the aim of delivering more value for the building occupants. For instance, SBD, BIM, TVD, and Design Quality Indicators can be used to better coordinate design teams and manage the design process so that communication-based inefficiencies can be avoided, and design change orders during construction can be reduced. The LPS has already been successfully used as an alternative to the CPM to better address variability during the execution of a project. Additionally, contractual mechanisms, such as the ones supported by IPD, can help develop more collaborative relationships between designers and contractors, where each of the agents within a project operates to support and reinforce the others, rather than operating competitively to one other. Lastly, Choose by Advantages (CBA), the Design Structure Matrix (DSM), Concurrent Engineering (CE), and the Benefits Realization Model (BRM/BeReal) can help to better structure and establish such relationships.

The missing pieces for establishing lean construction in the 4.0 era

We suggest that methods originating from lean construction can serve as a guiding compass for successfully implementing Construction 4.0. For example, the widespread use of BIM in the last decade has already shown that typical tasks currently executed by project managers, such as estimating and scheduling, are becoming automated, but have also revealed new ways of looking at project data more holistically, which can facilitate decision-making (Thomas & Bowman, 2021). The CBA method can provide a good framework for orienting decision-making toward the right direction, while TVD can serve as an alternative technique to costing by correlating costs with market-defined prices and expected financial returns. Existing BIM software platforms are mature in the sense that they provide at least four of the following features:

a. Integration of the various discipline models (such as architectural, structural, and Mechanical, Electrical, and Plumbing Equipment [MEP]) in a federated BIM model to gain a better understanding of the design;
b. Clash detection – analyzing clashes between building systems for constructability analysis;
c. Construction sequencing – linking of the project plan (mostly at the master plan level) to the model and creating a simulation of the project (also known as 4D CAD);
d. Design review and communication, including marking-up of designs for clarification;
e. Visualization of design during construction (i.e., viewing product information in the field);
f. Quantity and cost take-offs;
g. Constraints analysis;
h. Evaluation of "what-if" scenarios; and
i. Visual tracking of construction progress.

However, none of the existing software systems supports "pull" production management, such as a functionality that allows users to halt a specific process if it is identified as non-value-adding. Additionally, although BIM software solutions are now capable of: a) tackling planning issues from a high level to detailed daily tasks, and b) achieving top-down integration of product definition (i.e., a detailed 3D model) and construction planning (i.e., CM software), the bottom-up information flow from the site still remains a challenge. Thus, an opportunity arises to create integrated software systems that are based on lean construction methods, such as the LPS, to support trades personnel in their day-to-day work instead of just supporting top-down planning (lookahead/make ready processes). Although mobile technologies have improved communication between on-site crews, construction managers, and architects/engineers, they are mostly used in "view-only" mode by the practitioners on site. Such interfaces can be further implemented to monitor the process status in the future (e.g., scan-to-BIM processes) and to inform design and/or construction planning based on actual developments and input from the on-site crews.

Examples that hint toward that direction include the so-called KANBIM software systems, such as Dalux, VisiLean, and Ourplan, which are prototypical production workflow management software that integrate lean principles into BIM platforms. VisiLean is developed on the TFV framework, and the guiding idea is to explore how the representation of up-to-date process information in a BIM environment can influence the workflow itself, and how that could lead to improved decision-making related to waiting time and rework. It supports planning and scheduling at multiple levels (phase, lookahead, and weekly planning), task management using the last planner approach, as well as synchronous visualization of product and process. The project progress is updated as users in the field (primarily trade crew managers) report the as-made status of work packages using the interface provided (Dave et al., 2011). Additionally, materials, design information, equipment, and work itself can be "pulled" (rather than pushed) by the people on site using the same interface. VisiLean has been tested in laboratory and on-site settings that varied in scale, and some of the achieved efficiencies include workers' continuous access to up-to-date information, effective coordination between crews, and the reduction of informal communication. In addition, it significantly reduced the latency between actual progress tracking and plans (Tzortzopoulos et al., 2020).

Despite the advantages that were attainable via the implementation of such prototypes, further research is still needed to improve existing software systems to better integrate lean principles with Construction 4.0 technologies. Adding functionalities that allow capturing as-built data and providing in-process quality feedback is essential and is already considered in research. The idea of using augmented reality interfaces to help communicate work progress and methods and to further support decision-making is rather nascent and not fully exploited. The capacity of pulling detailing and fabrication/assembly information of building components according to short-term planning to match production flow could prove very beneficial for updating the product in a bottom-up way. Applying AI methods, such as machine learning, to automate product-based planning, planning optimization, and adjustment of short-term plans can help develop predictive plans and optimize production flow. Reducing the gaps in the information provided by software systems and the actual situation, between the designed and the as-built situation, can help build trust in the tools and make them more usable.

Finally, educating practitioners about the methods that lie behind the tools they use (e.g., the LPS) is considered crucial for convincing them to use such tools to overcome practical hurdles on site. The lack of understanding of how the tools work, in combination with an information gap between what is designed and planned in a digital environment and what is actually constructed, tends to cause frustration for workers on site and results in workers not using a new tool.

To summarize, traditional project planning and control methods focus on the product aspects, implementing activities from the viewpoint of the general contractor into a WBS, whereas trade subcontractors focus on the process aspects. Tools such as the ones mentioned above hold the promise of aligning objectives, both on a product and a process level as well as setting clear objectives on how to provide more value at the project level. The disruptive changes that have been brought about by Construction 4.0 may be significant, yet they may fall short of revolutionizing the AEC industry in a holistic fashion (similar to what happened in the 1990s). The introduction and application of a new set of guiding principles and techniques are necessary to organize such concepts and establish new paradigms for how we design, engineer, manage, and construct our built environment.

Discussion – How can lean thinking inform the research and education agenda of construction management in the 4.0 era?

The opportunity of applying lean thinking in construction has resurfaced recently due to the increasing rate of digitization and virtualization, not only in architecture and engineering but also in construction, which calls for a more systematic management of the flow of information. Additionally, the reinforcement of more stringent environmental policies on the built environment at the policy level requires the delivery of better-performing buildings with less variation in the promised deliverable quality. Below we highlight themes that can help frame solutions to existing challenges.

Early project collaboration should be considered more seriously and should be interpreted as a dialectical activity in which collaboration can be considered primarily as a social construct. Consequently, prescriptive and deterministic strategies and approaches should be avoided. Alternatively, the creation and utilization of project artifacts should emerge through collective reflection, building shared understanding among project participants. New forms of collaboration and multi-party contractual relationships, such as the ones supported by IPD, can reinforce this alignment by setting clear objectives on the project level, on the one hand, and promote the development of holistic design approaches, on the other hand. Introducing lean principles can help leverage new technologies that are prominent and becoming widely adopted across disciplines, such as BIM, AR/VR, and robotic construction, to name a few, and can help align actors who focus on products (i.e., architects, engineers, general contractors) and those who focus on processes (on-site trade contractors). Implementing methods such as the LPS in combination with BIM technologies can help actors to better manage both design and construction.

Misunderstandings about what represents collaboration in construction projects can emerge through the "clash" of different concepts of collaboration embedded in contracts, systems, and established paradigms. To overcome misunderstandings and be able to transform the industry, both researchers and practitioners need to work in unison and embrace the fact that the AEC is characterized by interactions that are socio-constructive in nature. The deep understanding of these interactions is of utmost importance when it comes to adopting new technologies and/or novel methods. Concepts of value generation and information flow can help us to better understand which technology is more appropriate, and to analyze the socio-constructive and dialectical nature of collaborative interactions at early project stages by examining how different concepts of collaboration coexist. Therefore, further empirical research should be developed to better explain how project participants create shared understandings using diverse lean tools (CBA, SBD, TVD, etc.) in early project stages. This can be achieved through the close interaction of academic and applied research.

Furthermore, the concept of continuous improvement cycles as an active and scientific learning process, which is almost absent in the traditional project-delivery methods but is central to lean thinking, is a very useful one. Beyond the tasks of organizing and planning, CM should also consider the aspects of adherence and continuous improvement. One of the main challenges that the application of lean construction is facing is AEC organizations' lack of initiative to learn to orient themselves toward continuous improvement. The notion of Total Quality Monitoring (TQM), which involves stopping production to resolve the origin of defects in the process as they arise, is almost non-existent in construction, and in its place, a "good enough to proceed" approach is used. Continuous improvement can be regarded as a research process that seeks to add value, whether it is considered a subset of design (i.e., "construction completes the design") or an initiative that directly contributes to the design of the product. The combination of these two approaches of continuous improvement can lead to a different paradigm that can also be referred to as "design science," a term introduced by Buckminster Fuller (1975) in the 1970s, and that has been also introduced into lean construction by Koskela.

Design science, as defined by Fuller, refers to the process of deliberately ordering components in the creation of new configurations (i.e., built structures) to achieve a desired goal. In this case, the focus is on supporting the process of creation by the effective employment of generalized principles derived by sound science. Design science incorporates several key concepts that are rarely integrated in human affairs, with the specific goal of helping the Earth's finite resources to meet the needs of all humanity without disrupting ecological processes. If employed effectively in the AEC, it can provide a coherent research framework for realizing a combination of economic, ethical, aesthetic, and superior performance in all design undertakings. The uniqueness of design science and its relevance with lean thinking was that it put great emphasis on the purpose of design value generation, and that it is generally applicable. The concepts of value generation and information flow as expressed in the 1990s similarly focused on eliminating value loss by analyzing requirements rigorously, systematically managing their flow-down, and seeking optimization. For instance, the concept of value generation has helped in understanding customers' perspectives (e.g., customer perceived value, customer personal values, customer satisfaction, value as an intersubjective phenomenon, value as the fulfillment of a purpose).

In summary, lean thinking can help reformulate the basis of both design and CM, as it offers a practical framework for processes of innovation and planning, and it is as relevant to product development or building design as it is to addressing the broader collective issues facing humanity and, specifically, the AEC in the 21st century. Lean thinking can become an enabler of industry 4.0 tools and concepts by offering standardized, transparent, and reproducible processes that are fundamental for the introduction of such innovations. Furthermore, it can help reduce product and process complexity, which can lead to the adoption of new technologies by ensuring the efficient and economic implementation of such tools in actual projects.

Research agenda toward Lean Construction 4.0

As the increasing rate of digitalization and virtualization is hinting toward an imminent and radical transformation of the AEC, what becomes critical is to realize how guiding principles and concepts are necessary to understand the changing requirements of the future and shape the industry accordingly, instead of being carried away by emerging technologies and tools. Since 2000, there has been a surge in the development of tools and technologies for the

AEC, but most of them have focused on automating tasks within the existing construction paradigm instead of questioning it and reframing it under the new circumstances. Yet, there are more pressing problems in digital project management today, such as the tracking of assembly delays and the monitoring of geometric deviances between the building design and the as-built product (i.e., TQM), which could be done in more efficient ways to improve project delivery. Having a better understanding of the specific problems in AEC (e.g., via root cause analysis) will help guide a more impactful path of technological development.

Future research should additionally focus on changing attitudes and behavior of all the actors involved in a project, as much as on introducing new tools and methods. Although new forms of contracts, such as the ones proposed by IPD, assume that collaboration is achieved through structural-organizational changes, limitations regarding the project managers' and designers' abilities to perform in such new situations may prevent their successful application. Enabling the early involvement of stakeholders and supporting a unified project culture play an important role for shifting paradigms, but the benefits of such structural changes should be understood and embraced by the people involved as well. Therefore, a deep understanding of the social fabric of the AEC in different regions is vital and is often overlooked in academic research (e.g., different rates of BIM adoption across regions). The technique of the "five whys of problem solving," which is the process of asking "why" five times whenever a problem is encountered, may prove useful for solving the root cause of problems (i.e., constant design changes) in the industry instead of just treating the symptoms (i.e., automating the process of updating design using parametric tools). In many aspects, the first attempts to introduce lean thinking in construction took a high-level approach and fell short on understanding the peculiarities of the industry at a socio-economic level; thus, the approach failed to become a widespread paradigm. This is because, in many cases, concepts were transplanted from other industries without considering the intricacies of the AEC, which vary greatly from region to region and, more importantly, because the introduction of new technology was often assumed to solve existing socio-economic issues. However, it has become clear that addressing the pressing environmental and socio-economic challenges (e.g., changing climate, political stability, etc.) has to place "people" at the center and account for AEC's impact at a global scale. This vision is also put forward by the recent agenda documents of the European Union on "Industry 5.0," with a special emphasis on sustainable, human-centric, and resilient industries (Cotta et al., 2021).

Undoubtedly, technology – and specifically software – has become an essential tool for assisting architects, engineers, and project managers alike in fulfilling their tasks; thus, the developments in technology and software have shaped their way of thinking. However, instead of focusing on what problem they need to solve, actors in the AEC often think based on what the software allows them to do (i.e., software-oriented thinking instead of problem-based thinking). For example, BIM tools were initially developed for automating the production of drawings, but over the years, they have become a project management tool that can be used throughout the life cycle of the project. The question that arises is, how would BIM and other design and management tools have looked if they were developed following lean principles (i.e., inventory, value, flow, process)? More importantly, how can lean thinking fit into established policies and the fragmented reality of the AEC, and how can it affect the development of tools and technologies as we move forward?

To answer these questions, academic research should focus on the very basis of how production is conceptualized, not only as the transformation of inputs into outputs, but also as a flow composed of transformation, inspection, moving, and waiting, as well as a process where value for the user is created through the fulfillment of requirements. Additionally, research should focus on how collaboration can be better defined and understood among people who have specialized functions of practice that are embodied in the different professional disciplines of AEC.

Conclusion

Given the digitization and automation trends in the industry and the associated innovation adoption paths of individual technologies and concepts, we have emphasized the need to "design the product with the process" and the early involvement of all stakeholders in design, principles that were shown to improve project outcomes from multiple angles.

The 4.0 framework combined with lean methods and techniques has the potential to enable the AEC industry to better focus on life-cycle outcomes, a perspective that has been completely missing and is probably one of the main reasons for the large carbon footprint resulting from the existing and new built environment. Additionally, the mentioned synergy can help overcome obstacles that hinder the application of innovative technologies (e.g., the highly fragmented nature of the industry). It can do so by refocusing the attention of all partners on production flow by supporting closer and more effective

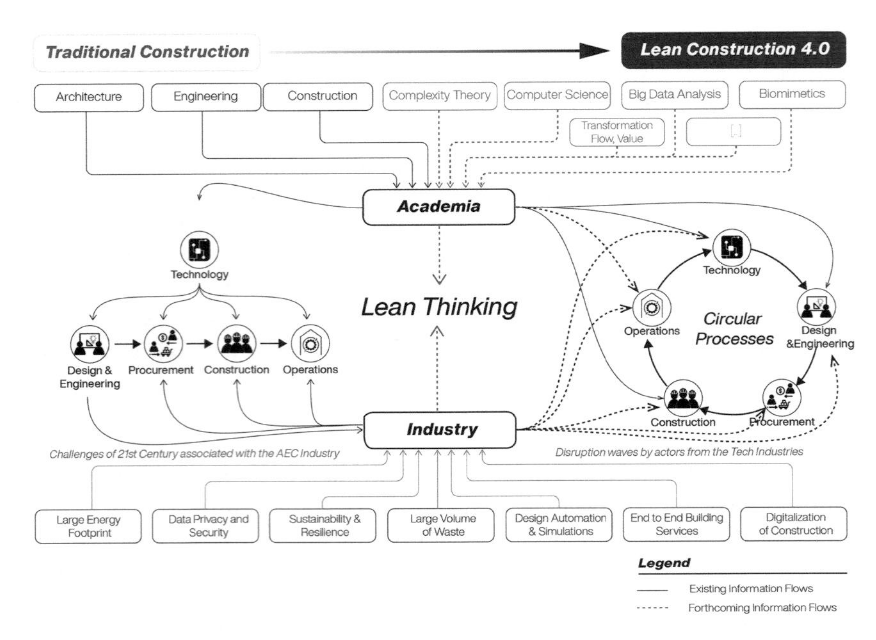

Figure 9.3 The flows of information between industry and academia across the life cycle of the building, and the change envisioned as the industry moves forward into the 4.0 era

collaboration between those who focus on product (i.e., general contractors) and those who focus on process (i.e., trade subcontractors).

From the viewpoint of integration, the literature on how design thinking needs to be transformed in a "lean manner" (i.e., seeking a unified design science beyond disciplinary fragmentation) represents a useful starting point in this endeavor. According to Uusitalo et al. (2017), this systemic view on design can be attained by the combination of different lean design management methods, tools, and techniques. An example is value generation/early design/optioneering, with links to client value, benefits, and BIM through the fast development of alternatives (Gerber et al., 2012; Malkawi, 2005). Other ideas include BIM and LPS implementation in design – specifically, investigating how they complement each other at detailed activity levels and how complexity theory could help build resilience in design management (Pantazis & Gerber, 2019; Suh, 2005).

On the design front, there is also fertile ground in utilizing more automation, data, and AI. Improved automation of design (e.g., via computational design approaches) could enable architects to evaluate more alternatives in terms of their performance against lean principles, tools, and methods to be applied in the project. This vein of thinking also deals with code checking, standardized products, platforms, mass customization, and so on. Naturally, these capabilities will be enabled only through educating a more computationally savvy design workforce, which connects to the earlier point on transforming academic programs in the AEC overall, and enriching them with methods and techniques related to complexity theory, biomimetics, and big data analytics, to name a few. By attending to these research themes in ways that build upon the seminal thinking that underpins lean thinking, the CM field can better address these issues.

Lastly, we find it essential to highlight that the mentioned trends and the vision for improvements in the AEC via the implementation of technology go through people, a parameter that is often neglected. In comparison with two or three decades ago, the technological options are endless today. However, to unlock the described synergy between technology and management of the built environment, we will first need to walk the academic research, followed by the industry, through a philosophical transformation in which the well-established lean construction literature could help. Figure 9.3 illustrates how this paradigm shift can happen by combining existing information flows with forthcoming ones, both on the academic and the industrial fronts, by placing lean thinking in the center and by embracing circular design processes.

References

Aoun, J. E. (2017). *Robot-proof: Higher education in the age of artificial intelligence.* MIT press.

Cotta, J., Breque, M., De Nul, L., & Petridis, A. (2021). *Industry 5.0: Towards a sustainable, human-centric and resilient European industry* (1st ed., Vol. 1). European Commission.

Dave, B., Boddy, S., & Koskela, L. (2011, July). *Visilean: Designing a production management system with lean and BIM* [Paper presentation]. The 19th Annual Conference of the International Group for Lean Construction, Lima, Peru.

Davis, D. (2021). Katerra's $2 billion legacy. *Architect Magazine.* https:// www .architectmagazine.com/technology/katerras-2-billion-legacy_o

Forcael, E., Ferrari, I., Opazo-Vega, A., & Pulido-Arcas, J. A. (2020). Construction 4.0: A literature review. *Sustainability, 12*(22), 9755.

Fuller, R. B. (1975). *Synergetics: Explorations in the geometry of thinking.* Estate of R. Buckminster Fuller.

Gao, S., & Low, S. P. (2014). Lean construction management. In S. Gao & S. P. Low (Eds.), *From lean production to lean construction* (pp. 27–48). Springer.

Gerber, D. J., Lin, S.-H. E., Pan, B. P., & Solmaz, A. S. (2012, March). *Design optioneering: Multi-disciplinary design optimization through parameterization, domain integration and automation of a genetic algorithm* [Paper presentation]. The SimAUD 2012, Orlando, FL.

Gordon, R. A., & Howell, J. E. (1959). *Higher education for business.* Columbia University Press.

Horman, M. J., & Kenley, R. (2005). Quantifying levels of wasted time in construction with meta-analysis. *Journal of Construction Engineering and Management, 131*(1), 52–61.

Koç, E., Pantazis, E., Soibelman, L., & Gerber, D. J. (2020). Emerging trends and research directions. In A. Sawhney, M. Riley, & J. Irizarry (Eds.), *Construction 4.0: An innovation platform for the built environment* (pp. 460–476). Routledge.

Koch, C. (2005). *Failures in combined knowledge and material supply chains* [Paper presentation]. The International Engineering Management Conference, St. John's, Newfoundland and Labrador, Canada.

Koskela, L. (1992). *Application of the new production philosophy to construction* (Vol. 72). Citeseer.

Koskela, L., Howell, G., Ballard, G., & Tommelein, I. (2002). The foundations of lean construction. In R. Best & G. De Valence (Eds.), *Design and construction: Building in value* (pp. 211–226). Routledge.

Koskela, L., Pikas, E., Niiranen, J., Ferrantelli, A., & Dave, B. (2017, 9–12 July). *On epistemology of construction engineering and management* [Paper presentation]. The 25th Annual Conference of the International Group for Lean Construction, Heraklion, Greece.

Levitt, T. (1960). Marketing myopia. *Harvard Business Review, 38*(4), 24–47.

Liker, J. K., & Meier, D. (2006). *Toyota way fieldbook.* McGraw-Hill Education.

Malkawi, A. (2005). Performance simulation: Research and tools. In B. Kolarevic & A. Malkawi (Eds.), *Performative architecture: Beyond instrumentality* (pp. 85–96). Spon Press.

Oesterreich, T. D., & Teuteberg, F. (2016). Understanding the implications of digitisation and automation in the context of Industry 4.0: A triangulation approach and elements of a research agenda for the construction industry. *Computers In Industry, 83*, 121–139.

Pantazis, E., & Gerber, D. J. (2019). Beyond geometric complexity: A critical review of complexity theory and how it relates to architecture engineering and construction. *Architectural Science Review, 62*(5), 371–388. https://doi.org/10.1080/00038628.2019.1659750

Sawhney, A., Riley, M., & Irizarry, J. (2020a). *Construction 4.0 : An innovation platform for the building environment* (Vol. 1). Routledge.

Sawhney, A., Riley, M., & Irizarry, J. (2020b). *Construction 4.0: An innovation platform for the built environment.* Routledge.

So, S., & Sun, H. (2010). Supplier integration strategy for lean manufacturing adoption in electronic-enabled supply chains. *Supply Chain Management: An International Journal, 15*(6), 474–487.

Sriram, C., Azam, M., & Nieuwland, M. v. (2015). *The construction productivity imperative.* McKinsey Productivity Sciences Center, McKinsey & Company Report. http://dln.jaipuria.ac.in:8080/jspui/bitstream/123456789/2501/1/The%20construction%20productivity%20imperative.pdf

Suh, N. P. (2005). Complexity in engineering. *CIRP Annals - Manufacturing Technology, 54*(2), 46–63. http://dx.doi.org/10.1016/S0007-8506(07)60019-5

Thomas, E., & Bowman, J. (2021). *Harnessing the data advantage in construction: Why adopting a data strategy can bring firms a competitive edge.* Autodesk & FMI Report. https://construction.autodesk.com/resources/guides/harnessing-data-advantage-in-construction/

Tommelein, I. D., & Ballard, G. (1999). Proceedings IGLC-7. In I. D. Tommelein (Ed.), *Proceedings of the Seventh Conference of the International Group for Lean Construction.* Lean Construction Institute.

Tzortzopoulos, P., Kagioglou, M., & Koskela, L. (2020). *Lean construction: Core concepts and new frontiers.* Routledge.

Uusitalo, P., Olivieri, H., Seppänen, O., Pikas, E., & Peltokorpi, A. (2017). *Review of lean design management: Processes, methods and technologies* [Paper presentation]. The Proceedings of the 25th annual conference of the International Group for Lean Construction, Heraklion, Greece.

Winter, M., & Szczepanek, T. (2008). Projects and programmes as value creation processes: A new perspective and some practical implications. *International Journal of Project Management, 26*(1), 95–103.

Womack, J. P., & Jones, D. T. (1997). Lean thinking—Banish waste and create wealth in your corporation. *Journal of the Operational Research Society, 48*(11), 1148.

10 Fast and frugal research: the pervasive use of questionnaires in construction management research

Dominic D. Ahiaga-Dagbui and Igor Martek

Introduction

Let's cut to the chase – we have an identity and credibility problem in construction management research (CMR). Many have laid the blame for our credibility problem at the feet of the fast and frugal methods we use for much of CMR, particularly the pervasiveness of perception-based questionnaires. It is not just a problem for Construction Management (CM) academics; our students – undergraduate to PhD – seem to reflect this as well. Much of our research fails to teach and lead the industry we are aligned with. We hardly push the boundaries of what is already known. A lot of what we do is almost akin to journalism – reporting what so-called industry 'experts' do or believe. We seek barriers, causes, important factors, limitations, success factors, and so on by sending out questionnaires to industry practitioners.

How did we get here? There are several possible answers to this simple question. We provide just two here. Firstly, we operate in a production-based publish-or-perish ecosystem that leads to research articles that should probably never have been written or published. What's the quickest way to get another publication? That's writ large. Perception-based questionnaires are fast and frugal. They might help to game the ill-conceived quantity-over-quality rewards system of universities, but ultimately, they will lead to irrelevance and a continuing waste of research resources.

Secondly, CM is a diverse field, with academic researchers approaching their studies from different disciplinary and methodological perspectives. From originally focussing on engineering and production-oriented research, CM has experienced a heavy influence of the social sciences since the 1980s. With this influence, theoretical traditions native to the social sciences became integrated with empirical construction research. The extensive use of perception-based surveys in CM research is best indexed by the proportion of articles in some of the leading journals reporting survey data. For example, at the time of writing (April 2021), we analysed the data sources of the articles published in the latest two issues of a prominent CM journal, *Construction Management Economics* (CME; 2021, Volume 39, Issues 6 and 7). Sixty per cent of these papers were based on surveys. In fact, 40 per cent of the most-cited papers in CME[1] are questionnaire-based papers about critical success factors. In addition, 50 per cent of the articles in the latest issue of *Engineering Construction and Architectural Management* (ECAM; 2021, Volume 28, Issue 7) were based on questionnaires or interviews. We also reviewed the data sources of the most recent Association of Researchers in Construction Management Conference (ARCOM 2020). Nearly 70 per cent of the papers published in the proceedings used a form of survey – interviews or questionnaires.[2]

From the social sciences came the idea that social conditions could be measured, assessed, or constructed. Survey research is a wide field that covers a variety of techniques that principally entail asking questions. A survey can range from a brief feedback form to extensive, in-depth interviews designed to elicit information about specific scenarios, events, or conditions. The assumption is that CM is a social science because it involves people; thus, asking them about their perceptions is not only valid, but necessary. This chapter will focus on the (mis)use of perception-based questionnaires in CMR. We will explore the theoretical and methodological issues associated with questionnaires – some of which go to the very heart of the (ir)relevance debate that dots the landscape of the CMR community. First, we will explore the nature of CM research in the next section.

Construction management – an island floating between the continents of engineering and management

CM is a curious field. Its difficulty in establishing credibility as a 'stand-alone' branch of knowledge has given rise to doubts regarding how the field should be advanced through academic research. Construction contributes between 6 per cent and 14 per cent of a country's gross domestic product (GDP), consumes as much as half of the world's resources, and is amongst the largest industries in terms of employment worldwide (Martek et al., 2018). Construction is of course critical to the economy, and the provider of the necessary infrastructure on which the economies of both developed and developing nations depend for growth. It is not ignored by politicians or the public, nor can it afford to be ignored by educators or academic researchers. However, while CM is both a manufacturing and service industry, as a research discipline, it tends to divide along the lines of either construction (civil) engineering, or management. In other words, the purely engineering aspects of CM research can expect to be captured by those properly qualified in engineering research – and these will be found in university engineering departments. On the other hand, pure management research will be undertaken by business research specialists from management faculties. Research pertinent to CM gets done, but not always – and not primarily – by those academics who style themselves as CM specialists.

What research niche, then, is left to the CM researcher? Horizontally, CM is represented by the intersection of engineering and management. But it can be compartmentalised vertically, as well. There are three layers: research at the technical level, the project level, and the company level (Lee et al., 2016). The technical level is concerned with performance aspects of building components, equipment, and processes. For example, how should exposed steel reinforcement bars be protected from corrosion? This kind of research has to be undertaken scientifically, through experimentation, modelling, trial-testing, verification, adapting, remodelling, re-testing, and re-verification (Gerwick, 1990). In short, this kind of research is pure performance engineering. Positivism is the prevailing ontology – data collection, verification, and analysis are everything.

At the other end of the spectrum, we have the research of the firm; here, the construction company itself becomes the object of investigation. Performance, too, is critical, but the lens through which performance is understood is sociological. That is, the firm is considered an independent, animate entity with goals, strategies, competitors, strengths, weaknesses, legal rights and obligations, and even moral attributes and 'will'. The predominant mode of investigation is case study research. The aim is to identify the plethora of forces and influences that converge on the firm and shape its behaviour. Theory building, thus, is what preoccupies the management researcher – given certain situations and certain aspirations, what best explains why firms do what they do? The prevailing ontology is constructivism (interpretivism) – which scholars use to understand the motivations driving enterprise actions.

To make this clear, consider the literature review section of any of the various academic papers – for example, those daring to examine why companies venture into new markets. Invariably, authors will invoke one theory or another as the medium by which the research question is pursued.

One can expect to find theories drawn from economics: organisational economics, transactional cost economics, internationalisation theory, or industrial-organisational theory. Or perhaps from production network theories: resource dependency theory, social network theory, institutional theory, cluster theory. Or maybe from value chain theories: a resource-based view of the firm, a knowledge-based view, a competence-based view, dynamic capabilities theory (Dikmen and Birgonul, 2006; McWilliam et al., 2020). The point is that, while concrete-pouring and market diversification may both be activities undertaken by construction firms, they have nothing whatsoever in common other than that fact. Researching these spheres of activity similarly have negligible commonality; their methodology, research approach, scale, actors, and data and outcomes, not to mention interested audiences, are all very different. This is why research on CM is effectively partitioned between engineering academics and management academics.

What, then, happens in the middle ground, at the project level? Since projects are the ubiquitous vehicle by which construction activity is conducted, this is where the CM researcher comes into their own. Specifically, it is the project nature of construction that distinguishes it from manufacturing or services; consequently, it is why CM lies outside the field of view of the common management literature. We often conceive construction projects as one-off 'events'. Each building, for example, is unique, even if only by virtue of being associated with a specific geographic location or different client. And this raises a further point – buildings cannot be moved; they are built where they will sit. In the oft-quoted words of Drewer (2001): '[In building construction] ... it is the factory that is mobile, and not the product' (p. 71). Thus, the bespoke nature of projects, their delivery characteristics (rather than ongoing fabrication runs), combined with the added challenges of setting up a factory on site prior to getting started on the building itself, all conspire to make building construction unique. These unique features, derived from the project nature of construction along with the difficulties, pitfalls, and concerns that result, are the bread and butter of CM researchers.

Figure 10.1 illustrates the divisions.[3] Building research can be partitioned into five streams: structural engineering, materials and technology, physical performance, geotechnology, and CM (Warszawski et al., 2007). All but the last are the preserve of engineering and of experimental, scientific research. Even within the fifth pillar, CM, there are themes that are properly engineering concerns. Then, topics such as construction economics, company management, and public management are well catered for by management research. What is left? 'Project management' – encompassing everything from risk, construction, waste, knowledge, and stakeholder management, as well as related knowledge areas, such as safety, estimating, procurement, and contracting. That is not to say that CM researchers limit themselves to project management alone; they don't. However, where they do venture beyond project management issues into, say, business economics, stakeholder management, or sustainable construction, they run up against tests of legitimacy. What does the CM researcher offer regarding construction labour pricing, say, that a labour economist cannot? The answer has to be 'context'. The CM researcher fishing for topics not exactly confined to building project delivery will borrow heavily from other fields, and apply them to the construction space. CM researchers take from other disciplines, not the other way around. Along these lines, Bröchner (2018) notes that:

> While CME authors often cite articles published in top economics journals, citations in the opposite direction simply do not exist … [I]t is unlikely that the mainstream reluctance to cite construction economists is due to an increasing irrelevance of topics associated with the construction industry. Instead, there remains a suspicion that academic quality is a problem. (p. 179)

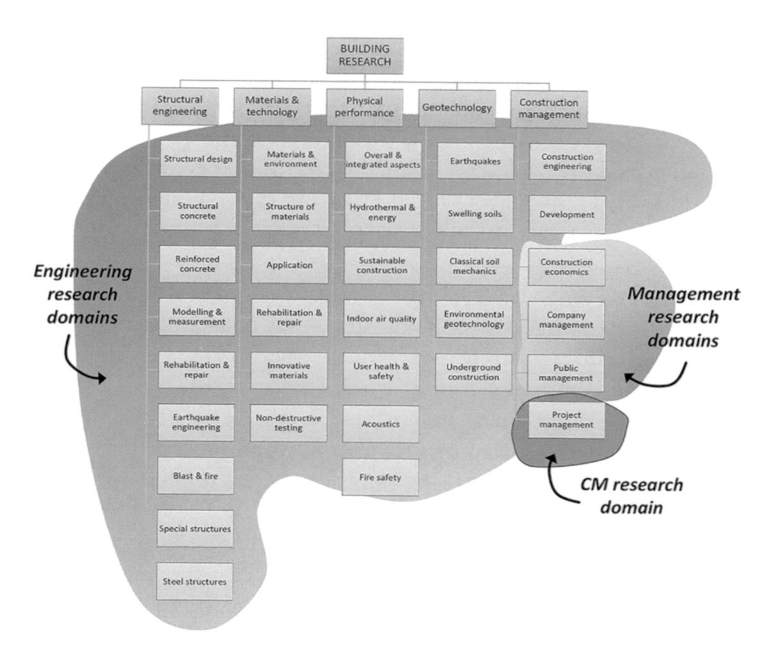

Figure 10.1 Building research fields are shared between the three disciplines of engineering, management, and construction management

The life and times of the CM researcher

Thus, we see how the CM researcher is constrained in the scope of their work. But how we go about our work is shaped by additional handicaps. Three factors are pertinent: the academic environment, industry engagement, and data availability.

Academic environment

CM, as an applied field, while relatively new, is today saturated with research scholars. In the postwar period, construction research was isolated within select engineering faculties. Through the 1970s, only a few universities worldwide recognised construction as distinct from engineering. Then, in the 1980s, construction 'took off'. Forty construction programmes were established in the USA at that time, with other fledgling programmes emerging in France, Germany, the UK, and Japan. By the turn of the century, there were over 100 in the USA and hundreds more worldwide, notably in China (Chinowsky and Diekmann, 2004; George et al., 2010). At the heart of the early debate as to the veracity of CM as a discipline worthy to be distinguished from engineering, was the question: 'whether or not construction engineering and management was a true academic pursuit and whether or not the field could develop research activities worthy of an academic degree or that could support professors embarking on a research and teaching career' (Chinowsky and Diekmann, 2004, p. 751).

For a while, the answer was 'yes.' A historical analysis of trends in CM research reveals shifts in research focus across the decades. Through the 1970s, research was preoccupied with capturing 'industry experience'. In the 1980s, it switched to measuring and improving project productivity. In the 1990s, the impact of technology on project management became the focus. Then, in the early 2000s, research shifted again to project planning and project life cycles. Since the turn of the century, however, the range of themes to be found in the CM literature has exploded exponentially. A search of Scopus, Web of Science, or Google Scholar will readily return 50 or so literature review papers dedicated to describing the research activity conducted on specific themes self-described as within the CM domain. In short, towards the end of the 1980s, CM research was a niche area conducted by a handful of specialists, mainly in engineering departments, on matters directly relevant to, and in collaboration with, the construction industry. They drew mainly upon operations research principles for their CM-related studies. Since then, the teaching of CM has burgeoned, spawning an abundance of schools populated with numerous academics with scant industry experience and limited contact with industry or awareness of their needs, but now running large doctoral programmes and being required to publish extensively to maintain professional credentials and career prospects (Willenbrock and Thomas, 2007; George et al., 2010; Chinowsky and Diekmann, 2004). Publications are, of course, key trading 'currencies' in academic circles. Ioannidis (2014) notes that these currencies are used to: 'purchase academic "goods" such as promotion and other power. Academic titles and power add further to the "wealth" of their possessor. ... This status quo can easily select for those who excel at gaming the system, producing prolifically mediocre and/or irreproducible research' (p. 3). This is one driver of fast and frugal research.

Industry engagement

As central as the construction industry is to the economy, it consistently generates low rates of profitability. In developed economies, that rate is on par, or just below, GDP growth – around 2–3 per cent. Thus, the industry is very much preoccupied with survival, and given the short-term project-by-project nature of its activities, it has very narrow concerns: building to schedule, building to quality, and building to budget. Of course, those concerns are merely doorways to a multitude of further issues – supply chain management, value engineering, innovative fabrication, and the like – but ultimately, it is fair to generalise that the industry may not be far-sighted, and it is usually absorbed by the problems of the day. To the extent that research is of relevance to construction, it is in facilitating the industry's two core activities: design and building. While the response of a CM researcher might well be, 'Great, I can help. Let's find out what others are doing, and let's learn from them,' this offers little assurance. Abstract solutions will have very little appeal when contractors are already embedded in a network of fellow industry practitioners from which they can extract working answers. Then again, if the question at hand is of a highly technical, ground-breaking nature, the problem set will be handed to the engineering department. They have the laboratory facilities, technical know-how, and scientific methodologies that CM researchers do not (Abudayyeh et al., 2004; Harris, 1992). The challenge from industry is one yet to be adequately met by CM researchers. Three decades ago, Robert Harris (1992) warned: 'The [construction] industry needs research to increase its competitive potential in innovative projects' (p. 431). We see the pressure on CM researchers to publish, in an environment overstocked with CM schools and academics, combined with a general disdain from industry as to the worth of their research efforts.

Data sources

There are generally four types of data available to us for research: observational, simulation, experimental, and derived. Observational data are typically collected via human observation, video and audio, surveys, or the use of instruments like sensors for monitoring and recording. Simulation data, on the other hand, may be generated by employing computer test models to simulate the functioning of a real-world process or system over time. Examples might include simulation of on-site crane-lifting operations, or fire response behaviour of on-site workers. The goal of simulation is to try to determine what would, or could, happen under certain conditions. Experimental data are collected through active intervention by the researcher to produce and measure change or to create a difference when a variable is altered. Experimental data typically allow the researcher to determine a causal relationship, and are generalisable to a larger population. Derived data are created by transforming existing data points, frequently from multiple data sources, using some type of transformation, such as an arithmetic formula or aggregation. An example might be population density data derived by integrating geographical area and population data. There are now numerous institutions, consultancies, and commercial data providers that service the 'research needs' of construction firms. IHS Global Insight, Business Monitor International, BIS Oxford Economics, Standard and Poor's, IBIS World, and Engineering News-Record are among many specialist agencies that inform companies on matters that concern them (Kang-Wook and Han, 2017).

As can be seen, there is a rather wide range of data sources that may be relied upon for CMR. Tellingly, CM researchers rarely resort to or exploit this diversity of sources. Yes, some of these do cost a good deal of money, and their full services are not covered by university library budgets. This is where the fast and frugal becomes rather appealing. Some might even argue that it is pragmatic – the only way. We thus go to academic journals for inspiration on what to write about. We use academic articles as templates for our work. We look for ways to carry out research that involves low-cost, easily obtainable, readily analysable material and data. We invariably look for 'research gaps' derived from the literature, rather than from industry. We look for earlier 'research limitations' as opportunities to extend on incomplete projects – the same study, but in a new country, or with added factors, or by asking different stakeholders. And we look for convenient research methods, the fast and frugal. Interviewing people is easier and cheaper than doing experiments, building prototypes, or measuring performance. Questionnaires are cheaper, quicker, and easier, still. They are the CM researcher's 'low-hanging fruit'.

The use of questionnaires

Surveys involve the use of a collection of questions, the survey instrument, to measure a construct that cannot be measured or observed directly. Likert scales are employed on the assumption that people's perceptions or opinions vary on a bi-polar scale, from negative (e.g. 'strongly disagree') to positive (e.g. 'strongly agree'). Survey respondents generally go through four stages of the response process when answering survey questions: 1) they must first understand the question, 2) then retrieve relevant information, 3) integrate the information to form a judgement, and finally 4) map their response to the available response options (Tourangeau et al., 2000). Typically, the items that inform questionnaires are generated from a combination of a literature review and/or focus groups. In this process, a list is developed having a common 'quality', such as 'barriers', 'goals', 'risks', or 'importance'. However, unlike other disciplines, the lists are rarely drafted to test or develop theory. Consequently, the best that can be derived from such studies is a rank order of relative importance or correlation.

When conducting structured surveys, a formula is employed to estimate the true value of the construct by combining the responses of the respondents to the questions. This scoring is typically undertaken electronically to generate the scale score or relative importance index. The assumption, therefore, is that for any respondent who answers a question in the survey instrument, it is possible to obtain an estimated value of the unobservable construct. Statistical analysis is then used to evaluate the responses to generate multiple scales that may be independent or correlated, as well as to perform tests of reliability, validity, and sensitivity.

This sort of research is thus particularly useful and appropriate where the only source of information needed is the subjective views of respondents, for example, in clinical research where the measurement of health interventions and outcomes, such as pain, distress, well-being, quality of life, disease-specific symptoms, or treatment-related side effects, are to be assessed (Testa and Simonson, 2017). Sample surveys are fast and frugal: a low-cost and convenient approach to data collection in comparison to other forms of data collection (e.g. lab experimentation). They can be administered globally and very quickly via email, social media, and websites. In a very short time, it is possible to acquire a large amount of information from a lot of people. In fact, you can even rely on for-fee survey services online to gather and perform all the associated statistical analyses for you. You need almost no understanding of the fundamental assumptions underlying the statistical analysis and results. It is no surprise that this tool is attractive to many researchers and students.

Questionnaire use, relevance, and impact

However, perception-based methods are known to be incompatible with research contexts and problems that are messy and complex, and where decisions are often made with incomplete and incoherent data. In fact, Azhar et al. (2010) claim that one of the main reasons why construction research does not seem to make a ready impact on the industry it is associated with is because of the predominant use of surveys. They argue that 'academic research loses relevance and has little or no impact on practice. One reason for this unfortunate state is that research methods used by academics in [Construction Engineering Management], namely, surveys and case studies, have mostly studied phenomena that has already occurred, i.e., it has focused on existing reality' (p. 87). Now, there is nothing inherently wrong with attempting to deconstruct and study what has happened in the past, especially if knowledge can be generated to shape the future. But, alas, this is typically not the goal of many of the surveys used in CMR. In their first editorial as the new editors of CME, Leiringer and Dainty (2017, p. 2) echoed similar sentiments:

> In recent years, we have seen a marked decline in empirical research and papers that observe, measure and study practice. Instead, we have seen a proliferation of papers where the research is seeking the opinions of respondents, typically through surveys or interviews. And so what we know about construction is largely framed by what practitioners tell us about it, and not what we observe in and through our engagements. (p. 1)

In most cases, what is to be derived from survey research has limited applicability in the real world. This limitation is underlined by the fact that such studies ask practitioners about what they do or believe. Consequently, insights cannot move much beyond the boundaries of what is already known. The researcher learns about the industry, but cannot teach or lead the industry. In a response to the question, 'Has academic construction management research been able to influence the practice?', Koskela (2017) argued that 'Both fields, project management and CM, have a history of low performance against expectations. However, progress in these fields has mostly been triggered by innovation engendered in practice; it is difficult to pinpoint results from academic research with major relevance and impact' (p. 15).

Limitations of the questionnaire approach

Beyond the relevance and impact problem, there are further issues that may be direr – questionnaires are fraught with many problems, limitations, and assumptions, particularly when theory generation or significant contribution to the knowledge or practice is the goal.

Validity, reliability, and response bias

The notions of reliability and validity are fundamental and critical in psychometric measurements. They are used to assess the accuracy of the measurement scales deployed for the research. Our working definition of reliability is the degree to which a measurement is the same each time it is undertaken. It gives an idea of the stability or consistency of the measurement instrument and the extent to which a score is free of random error. In other words, it helps to measure the error associated with the measurement. A more technical definition provided by Testa and Simonson (2017) is that reliability assesses the ability of the scale to remain stable during a period when external influencing factors are negligible (i.e. if most external factors remain fairly similar, the responses generated should not differ). An example of this in the physical sense is that a concrete column purporting to be 2.5 m high will always be that height every time it is measured. A closely related concept to reliability is internal consistency – how well different items or questions measure the same *construct*. Ideally, there should be a reasonably high level of consistency between different questions measuring the same construct (i.e. if the 'same' question is asked in two different ways, they should elicit similar responses). The Cronbach's alpha test is probably the most common consistency statistic.

A reliable score does not mean the results are necessarily valid, only that the measuring score is stable. Validity, on the other hand, is concerned with the meaning and interpretation of a scale used. While there are several dimensions to validity, including content, criterion, and construct validity, we will focus on the latter only for the sake of brevity. Constructs, of course, are theoretical abstractions developed to conceptualise a particular phenomenon. Thus, construct validity refers to the degree to which the measuring scale correlates with the construct being studied. This is fundamental, especially for generalisations, as it gives an indication of the legitimacy of the inferences that can be made from the scores. In other words, it reflects the strength of the conclusions you can reliably make, a simultaneous process of measure and theory validation. Validity coefficients are commonly seen as convincing proof of the validity or reliability of a particular procedure.

Fundamentally, humans – consciously or otherwise – synthesise numerous sources of information to develop an appropriate response in a given situation. Our responses are thus influenced by many factors, such as our interest in the subject matter, the phrasing of questions in surveys, whether the survey is in our first language or not, the length of the survey, and so on. In fact, in some cases, we respond by simply providing what we assume is the socially desirable answer. As far back as 1983, Rossi et al. observed that 'it became increasingly clear (and alarming) that unreliability in single-item measures was a major factor in producing results that were simply uninterpretable' (p. 16).

Memory and recall also play a crucial role in the reliability of survey results. It would come as no surprise that we humans do not store up facts and events in an easily accessible manner. Just and Carpenter (1992) thus found that individuals differ markedly in how much information they can hold in their working memory; therefore, surveys that overload the memory capacity of respondents elicit poor answers. Wikman (2007) writes:

> It is certainly not an easy task to recall a lot of facts and handle them in a stringent fashion. This may only be possible if respondents have encountered the same questions many times before, and what is asked for is very important and well-known to them... This makes it really difficult to know whether differences between samples or points in time are real differences or if they are only due to measure artefacts. (p. 27)

Tourangeau (2020) also noted that:

> The wide range of beliefs and values that respondents might consider in formulating their answers and the small number they actually considered suggests that variability in their answers over time could reflect unreliability in the processes of retrieving considerations and combining them into the relevant judgement. To the extent that respondents retrieve different considerations on different occasions, evaluate them differently, or use different methods to combine them, their answers to the same attitude question will vary over time. (p. 3)

The last part of the quote above should probably make one think twice, and carefully, about using surveys. However, beyond memory and recall lie a whole range of further fragilities collectively termed response bias. This is a general term used to describe a range of the tendencies for participants to respond inaccurately or falsely to questions (Mazor et al., 2002). They include *social desirability bias* (providing what might be socially acceptable responses); *non-response bias* (non-representative samples); *order effects* (order of questions); *recency bias* (cognitive bias favouring recent events, placing more emphasis on what comes to mind readily); or *extreme response bias* (respondents providing either the extreme negative or positive response).

Compounding this concern is that surveys with response bias frequently have a high level of statistical reliability, leading the researcher into a false sense of confidence regarding the validity of the measured constructs in the survey.

For example, studies have shown that the characteristics of the questionnaire itself can have a significant influence on the reliability of the responses. In some cases, seemingly insignificant and minor issues influence the type of response received. A few of these are listed below:

1. Questions about facts generally lead to more reliable answers than those based on non-factual or opinion-based questions (Hout and Hastings, 2016);
2. The question's location in the questionnaire – questions that appear later in the questionnaire typically generate more reliable responses (Saris and Gallhofer, 2007);
3. The type of scale utilised – Item-specific response scales produce more dependable answers (Wikman, 2006; Revilla et al., 2014);
4. False presuppositions about an event can cause responders to misremember the event as if the presuppositions were correct (Loftus, 1979);
5. Questions that do not include a middle response choice yield more reliable responses (Saris and Gallhofer, 2007);
6. Lengthy or complex questions increase the mental load placed on respondents, leading to misinterpretations (Just and Carpenter, 1992);
7. Questions about topical or salient subjects produce more reliable answers than questions about less significant topics (Saris and Gallhofer, 2007);
8. The number of responses and number of questions also affect the reliability of the results: fewer response alternatives and fewer questions result in more reliable responses (Revilla et al., 2014);
9. Attitudinal or behavioural questions result in increased rates of problematic substantive and non-substantive answers relative to demographic questions (Fowler, 2011); and
10. Questions about the past provide more reliable responses than those about the present and future (Saris and Gallhofer, 2007).

Beyond the issue of the reliability of the responses is another that we might be all too familiar with – respondent fatigue. We almost assume that survey respondents will give careful thought, time, and attention to the questionnaires we send out. Rokeach (1968) famously wrote, 'public opinion towards surveys amounts to no more than unstable surface opinions, rather than definable attitudes in any psychological sense' (p. 455).

Survey designers hope that respondents will provide carefully thought out and honest responses. But, let's face it, who wants to complete another survey? Do you wonder why response rates are frustratingly poor? For those who complete the surveys, reaction time studies have shown that people respond to the majority of attitude questions in roughly three seconds (Bassili and Scott, 1996). In some cases, respondents choose their answers before fully reading the questions. So much for carefully collected data, right? So, how are we to approach or report research results from surveys, when seemingly incidental tweaks of the questions can completely change the results and conclusions that can be derived? With a good dose of caution and circumspection.

Correlations and causation

One of the weaknesses of survey research is the oversimplification of reality. Rather than provide evidence of causation, correlational analysis (e.g. multiple regression and structural equation modelling) associated with surveys merely indicate some sort of association between variables. If we find that a variable A is correlated with B, it could simply mean that A caused B, or B caused A, or certain confounding variables caused both A and B without there being any causal relationship between A and B at all.

Ubani et al. (2013) investigated the causes of cost and schedule overruns in Nigeria and developed a questionnaire based on '110 hypothetical cost overrun' factors derived from the literature. Responses were evaluated by deriving the relative importance of factors and correlation coefficients. They found that material-related issues, including price fluctuations and shortages, were the main causes of overrun. They rejected the hypothesis that contractual relationships, labour, and design had any significant influence on cost overrun. They then recommended that clients, contractors, and consultants 'pay more attention to both material and external factors for there to be effective and efficient delivery on construction projects at the right time and cost' (p. 73). Several cost overrun studies have been conducted in this manner, albeit in different contexts (Ameh et al., 2010; Durdyev et al., 2012; Memon et al., 2012; Kaming et al., 1997). Ahiaga-Dagbui et al. (2017) examined the methodological weaknesses in the dominant approaches adopted to explain cost overrun causation on infrastructure projects and concluded that much of the research in this area is 'superficial, replicative, and thus has stagnated the development of a robust theory to mitigate and contain the problem' (p. 88). One does not have to search too hard to see the same approach being applied to many other areas of construction research in search of 'barriers', 'challenges', 'critical success factors', 'causes', and the like. The problem is that surveys cannot give definitive evidence of cause and effect.

The approach described above is typically based on correlation analysis in multiple linear regression. Independent variables are assumed to be analytically separable causes of the outcomes under investigation. However, multiple linear regression is only suitable for problems with a small number of variables, well-established causal relationships (such as elasticity of income or price), and where a large amount of reliable and valid data exists. However, where significant complexities and interdependency exist between variables, regression models tend to return rather spurious results. One of our favourite examples to illustrate this problem is as follows. Using stepwise regression, a study began with 31 observations and 30 potential variables (only variables with a *t* value greater than 2.0 are included). The adjusted R-square value was 0.85 with eight significant predictive variables. This would seem to be a model with good fit on perfunctory examination, until it is revealed that the original data were from a book of *random numbers* (Armstrong, 1970)! Statistical manipulation, as well as misapplication, has lulled us into an illusion of robustness, resulting in many inferences and conclusions that cannot be substantiated. Armstrong (2012) further adds: 'Analysts assume that models with a better fit provide more accurate forecast. This ignores the research showing that fit bears little relationship to ex-ante forecast accuracy. Typically, fit improves as complexity increases, while ex-ante forecast accuracy decreases' (p. 691).

Most of us have heard the phrase 'correlation is not causation'. Rather than provide evidence of causation, correlations merely indicate some sort of association between variables. Using cost overrun causation as an example, Ahiaga-Dagbui et al. (2017) observed that many survey studies have attempted to identify the root causes of cost overruns, but invariably only scratch the surface of this complicated problem using statistical measures of correlation between variables. They argued that a strong correlation only provides 'circumstantial evidence' implying a plausible causal link. As pointed out by Ahiaga-Dagbui et al. (2017), correlational analysis assumes that the presence of a given plausible causal condition will lead to the occurrence of the outcome. It implicitly presumes *causal asymmetry* as well: that the absence of a given cause or set of causal conditions thought to be associated with the outcome will result in the absence of the outcome. This assumption is usually faulty for a lot of complex problems, as there are often several possible causal paths for the problem (*equifinality*). For systemic and complex problems, those embedded in socio-technical, political, and economic environments, our approach should preferably move from independent, single-cause identification and traditional net-effect correlational analysis to look for *multiple conjunctural causation* (Ragin, 2004). Our methods have to carefully consider high-level interactions between multiple factors, multiple feedback loops, and the highly dynamic context of projects. As Ahiaga-Dagbui et al. (2017) argue in the case of cost overrun research, the crucial skill in understanding causation is not the ability to list or rank factors, but the capacity to analyse connections, interactions, and plausible causal combinations.

Contextual and semantic effects

The way questions are used in surveys may give the appearance that they are designed based on what is of the most theoretical interest to the researchers, regardless of whether or not they are technically difficult to quantify. For example, we assume that the meaning of words, or their interpretation by respondents, would be relatively stable and immune to changes. And yet, this assumption is hardly ever the case. It is fairly common for respondents to struggle to understand the meaning of some questions, even though the language appears to be straightforward to those who wrote them (Tourangeau et al., 2000). For non-native speakers of the survey language, questions that are complicated and place a larger cognitive strain on the respondent will generally generate less reliable responses. Language issues such as complex syntax and lexical items, for example, questions containing colloquial phrases or less commonly used terminology, may be particularly difficult to understand, thus skewing results (Park et al., 2016). Worse still, even frequently used words, such as 'often', 'sometimes', 'usually', 'seldom', and 'important', are ambiguous (have multiple meanings) or vague (have imprecise ranges of application). Varying meanings might be due to the context of usage, organisation, circumstances, and viewpoint of the respondent. Construction terminologies, such as 'labour forces', 'cost overrun', 'contingency', 'quality', 'scope', 'budget', and 'escalation', can take on a different meaning depending on the respondent, leading to steady construct validity issues. When respondents misinterpret even well-formulated questions, the questions they intend to answer may not be the ones that the researchers intended to ask in the first place.

Poor project management, lack of coordination between parties, mistakes during construction, and slow information flow between parties are some of the factors used in the survey by Memon et al. (2012) investigating cost overrun causation. Other factors, including inadequate control procedures, slow decision-making, waiting for information, and poor documentation, used in Frimpong et al. (2003) are rather too ambiguous. They may also evoke countless possible scenarios and examples depending on the context. This may be an indication that such factors are rather too superficial and therefore must be broken down further. It also follows that when questionnaire respondents are not thinking within the framework of the same or similar projects, their frames of reference can significantly differ, unless they were perhaps used as part of a structured-case study (Ahiaga-Dagbui et al., 2017).

Good surveys rely on asking carefully thought-out questions. A good question, for example, should be easy to understand, succinct, and not double-barrelled or biased. Even when questions are appropriately formulated and well-intentioned, they can turn out to be insufficient or even irrelevant in light of the respondents' cultural and value systems. Studies have demonstrated that respondents from different cultural or ethnic origins differ in their understanding and interpretation of the same survey questions (Warnecke et al., 1997). Respondents with a higher level of education also offer more reliable responses (Alwin and Krosnick, 1991). Unlike in an interview, questionnaire respondents cannot usually ask for clarifications and thus might often complete the survey with a faulty understanding of what the key constructs are. This defeats one of the key assumptions in surveys: that respondents are well-informed and clearly understand the questions being asked. There is no way of knowing if the respondent actually understood the content and questions. We do not know whether they even read the question thoroughly or why they skipped a particular question.

When there is a breakdown in communication during the survey process, the data will skew in unexpected directions. In other words, what you might presume to be excellent results from a very robust analysis of responses could well be a fluke. Some would argue the use of statistical tests to show that their findings are robust or 'significant'. However, the so-called statically significant relationships identified between constructs or measurement items are often 'erroneous or overstated', according to the analysis undertaken by Macleod et al. (2014). Writing about some of the statistical methods associated with surveys, Bennis and O'Toole (2005) observed that 'statistical and methodological wizardry can blind rather than illuminate' (p. 99). In fact, under conditions of severe reporting bias, statistical significance has nearly lost its capacity to distinguish between valid research and that which is erroneous and exaggerated. It is no surprise that it is incredibly difficult to replicate the results of many survey-based studies.

Conclusions

We have established the ubiquitousness of questionnaire surveys in CMR and detailed some of the associated problems with their use. Does this over-reliance point to an underlying crisis fermenting below the surface of the CM discipline? The diagnosis is not positive; the prognosis is worse. In essence, questionnaire-based research does little to promote the advancement of the practice and knowledge of CM. Findings deliver little to no practical benefit to the CM community, but instead are superficial and often replicative – a mechanism that merely delivers material to sustain journal publications. Consequently, the construction industry does not really read CM research journals – at least not in the way that medical practitioners rely on medical journals to update state-of-the-art practice. The main reading audience for CMR journals is CM researchers. Largely divorced from the day-to-day drivers, operations, and concerns of the construction industry, we read CM papers, looking for low-hanging 'research gaps' that will allow us to add yet another fast and frugal publication to our CVs. We perpetuate the vicious cycle that appears notoriously not self-correcting.

To be fair, questionnaires are a well-established and respected research method that is frequently used in psychology and the social sciences. These fields, of course, are mainly preoccupied with human behaviour, and as such seek to understand the relationship between human cognition of the world and the resulting human interactions between people and their environment. To the extent that CMR might seek to understand human perceptions of CM-related phenomena, questionnaires may serve well. But such pursuits are about people within CM, not CM itself. To suppose that cognitive perceptions can tell us anything substantive about advancing the CM discipline or practice is the first and overriding deficiency of the questionnaire approach. Beyond that fundamental criticism, questionnaire methodologies are subject to a range of fragilities that limit their 'truth-mining' potential. Memory inaccuracies, over-simplification, construct effects, poor questionnaire design and distribution, misunderstanding of questions, along with respondent fatigue all conspire to erode the value of the data extracted.

Beyond surveys, there are many untapped sources of data that could be used in CMR – experimental and simulation data; video recordings (e.g. for on-site safety behaviour studies); Internet of Things (IoT) sensor data; data from wearable technologies (e.g. physiological and cognitive measurements using wearable sensors); social media data; company reports; incident reports; print media; and so on. Free (unstructured) text data can be analysed using text-mining approaches like natural language processing. Natural language processing algorithms can evaluate enormous amounts of unstructured text-based data and are able to uncover relationships and propositions that would otherwise stay hidden in the vast amount of textual data. Furthermore, there are many databases that archive data that could be used for construction research – for example, Google's Dataset Search[4] and Public Data Explorer,[5] DataHub,[6] and UNdata,[7] to name a few. The point here is that there is an extensive gamut of data sources to draw from to support CMR – why rely on a very limited, easily skewed, and largely problematic data diet? We earlier detailed that nearly 70 per cent of the 2020 ARCOM conference papers and 60 per cent of the last two issues of CME articles are survey-based – this dominance should probably make us all worried.

Why, then, do we seem to be stuck in this rut? Why have surveys become the method of choice for most CM researchers? Perhaps we assume that what people tell us is the main source of 'truth' or 'reality'? Or perhaps we have assumed that a survey of a dozen or so construction managers or architects serves as a proxy for the mindset, belief, or practice of such groups? Or maybe we presume that the people we survey have *the answers* to the theoretical problems under investigation? Maybe the fast and frugal serves a more strategic goal for many – gaming the ill-conceived metric-driven university system. Surely, it's not a case of an inability to innovate, learn from other disciplines, or work closely with industry to influence or shape it in a manner that makes our research endeavours worthwhile. We dare to rule out, perhaps naively, the possibility that we really do not care much for robust theory development or making real contributions to the extant knowledge or practice. Whatever the reasons might be, the writing is on the wall: we either innovate and pursue research that can make profound contributions to practice and knowledge, or we lose any remaining credibility we might still have as a research community.

Notes

1. As of September 2021, see: https://www.tandfonline.com/action/showMostCitedArticles?journalCode=rcme20.
2. ARCOM is arguably the largest global construction management-focused conference. Since 2004, ARCOM conferences have attracted delegates from 47 countries (see: http://www.arcom.ac.uk).
3. We appreciate that the traditional construction management field as presented in Figure 10.1 has changed over the years. For example, construction information technology (BIM, digital construction, etc.) and construction law research have become core parts of construction management research, although surveys are not as popular in these areas.
4. See: https://datasetsearch.research.google.com/.
5. See: https://www.google.com/publicdata/directory.
6. See: https://datahub.io/search.
7. See: http://data.un.org/Default.aspx.

References

Abudayyeh O, Dibert-DeYoung A and Jaselskis E (2004) Analysis of trends in construction research: 1985–2002. *Journal of Construction Engineering and Management* 130(3): 433–9.

Ahiaga-Dagbui DD, Love PED, Smith SD, et al. (2017) Toward a systemic view to cost overrun causation in infrastructure projects: a review and implications for research. *Project Management Journal* 48(2): 88–98.

Alwin DF and Kro Krishnamurthysnick JA (1991) The reliability of survey attitude measurement: the influence of question and respondent attributes. *Sociological Methods and Research* 20(1): 139–81.

Ameh OJ, Soyingbe AA and Odusami KT (2010) Significant factors causing cost overruns in telecommunication projects in Nigeria. *Journal of Construction in Developing Countries* 15(2): 49–67.

Armstrong JS (1970) How to avoid exploratory research. *Journal of Advertising Research* 10(4): 27–30.

Armstrong JS (2012) Illusions in regression analysis. *International Journal of Forecasting* 28: 689–94.

Azhar S, Ahmad I and Sein MK (2010) Action Research as a proactive research method for construction engineering and management. *Journal of Construction Engineering and Management* 136(1): 87–98.

Bassili JN and Scott BS (1996) Response latency as a signal to question problems in survey research. *Public Opinion Quarterly* 60(3): 390–9.

Bennis WG and O'Toole J (2005, May) How business schools lost their way. *Harvard Business Review*. https://hbr.org/2005/05/how-business-schools-lost-their-way

Bröchner J (2018) Construction economics and economics journals. *Construction Management and Economics* 36(3): 175–80.

Chinowsky PS and Diekmann JE (2004) Construction engineering management educators: history and deteriorating community. *Journal of Construction Engineering and Management* 130(5): 751–8.

Dikmen I and Birgonul MT (2006) A review of international construction research: Ranko Bon's contribution. *Construction Management and Economics* 24(7): 725–33.

Drewer S (2001) A perspective of the international construction system. *Habitat International* 25(1): 69–79.

Durdyev S, Ismail S and Bakar NA (2012) Factors causing cost overruns in construction of residential projects: case study of Turkey. *International Journal of Science and Management* 1(1): 3–12.

Fowler FJ (2011) Coding the behavior of interviewers and respondents to evaluate survey questions. In: Madans J, Miller K, Maitland A, et al. (eds.) *Question evaluation methods: Contributing to the science of data quality*. Hoboken, NJ: Wiley, pp. 5–21.

Frimpong Y, Oluwoye J and Crawford L (2003) Causes of delay and cost overruns in construction of groundwater projects in a developing countries; Ghana as a case study. *International Journal of Project Management* 21(5): 321–6.

George W, Huanqing L and Zhaomin R (2010) Globalisation in construction management education. *Journal of Applied Research in Higher Education* 2(2): 52–62.

Gerwick Jr BC (1990) Implementing construction research. *Journal of Construction Engineering and Management* 116(4): 556–63.

Goldstein DG and Gigerenzer G (2009) Fast and frugal forecasting. *International Journal of Forecasting* 25(4): 760–72.

Harris RB (1992) A challenge for research. *Journal of Construction Engineering and Management* 118(3): 422–34.
Hout M and Hastings O (2016) Reliability of the core items in the General Social Survey: estimates from the three-wave panels, 2006–2014. *Sociological Science* 3: 971–1002.
Ioannidis JPA (2014) How to make more published research true. *PLoS Medicine* 11(10): e1001747.
Just MA and Carpenter PA (1992) A capacity theory of comprehension. *Psychological Review* 99: 122–49.
Kaming PF, Olomolaiye PO, Holt GD, et al. (1997) Factors influencing construction time and cost overruns on high-rise projects in Indonesia. *Construction Management and Economics* 15(1): 83–94.
Kang-Wook L and Han SH (2017) Quantitative analysis for country classification in the construction industry. *Journal of Management in Engineering* 33(4): 1.
Koskela L (2017) Why is management research irrelevant? *Construction Management and Economics* 35(1–2): 4–23.
Lee KW, Han SH, Park H, et al. (2016) Empirical analysis of host-country effects in the international construction market: an industry-level approach. *Journal of Construction Engineering and Management* 142(3): 04015092.
Leiringer R and Dainty A (2017) Construction management and economics: new directions. *Construction Management and Economics* 35(1–2): 1–3.
Loftus EF (1979) *Eyewitness testimony.* Cambridge, MA: Harvard University Press.
Macleod MR, Michie S, Roberts I, et al. (2014) Biomedical research: increasing value, reducing waste. *The Lancet* 383(9912): 101–104.
Martek I, Hosseini MR, Shrestha A, et al. (2018) The sustainability narrative in contemporary architecture: falling short of building a sustainable future. *Sustainability* 10(4): 981. https://doi.org/10.3390/su10040981
Mazor KM, Clauser BE, Field TS, et al. (2002) A demonstration of the impact of response bias on the results of patient satisfaction surveys. *Health Services Research* 37(5): 1403–17.
McWilliam SE, Kim JK, Mudambi R, et al. (2020) Global value chain governance: Intersections with international business. *Journal of World Business* 55(4). https://doi.org/10.1016/j.jwb.2019.101067
Memon AH, Rahman IA and Aziz AAA (2012) The cause factors of large project's cost overrun: a survey in the southern part of Peninsular Malaysia. *International Journal of Real Estate Studies* 7(2): 1–15.
Park H, Sha MM and Willis G (2016) Influence of English-language proficiency on the cognitive processing of survey questions. *Field Methods* 28(4): 415–30.
Ragin CC (2004) Turning the tables: How case-oriented research challenges. In: Brady, HE and Collier D (eds.) *Rethinking social inquiry: diverse tools, shared standards.* Lanham, MD: Rowman & Littlefield, pp. 123–38.
Revilla MA, Saris WE and Krosnick JA (2014) Choosing the number of categories in agree–disagree scales. *Sociological Methods & Research* 43(1): 73–97.
Rokeach M (1968) Attitudes. In: Sills DL (ed) *International encyclopedia of the social sciences.* New York: Macmillan and Free Press, pp. 449–57.
Rossi PH, Wright JD and Anderson AB (1983) Sample surveys: history, current practice, and future prospects. In: Rossi PH, Wright JD and Anderson AB (eds) *Handbook of survey research.* New York: Elsevier Science & Technology, pp. 1–20.
Saris WE and Gallhofer I (2007) Estimation of the effects of measurement characteristics on the quality of survey questions. *Survey Research Methods* 1(1): 29–43.

Testa MA and Simonson DC (2017) The use of questionnaires and surveys. In: Robertson D and Williams GH (eds) *Clinical and translational science* (2nd ed). Amsterdam: Elsevier, pp. 207–26.
Tourangeau R (2020) Survey reliability: models, methods, and findings. *Journal of Survey Statistics and Methodology* 9(5): 961–91. https://doi.org/10.1093/jssam/smaa021
Tourangeau R, Rips LJ and Rasinski K (2000) *The psychology of survey responses.* New York: Cambridge University Press.
Ubani EC, Okoroch KA and Emeribe SC (2013) Analysis of factors influencing time and cost overruns on construction projects in South Eastern Nigerian. *International Journal of Management Sciences and Business Research* 2(2): 73–84.
Warnecke RB, Johnson TP, Chávez N, et al. (1997) Improving question wording in surveys of culturally diverse populations. *Annals of Epidemiology* 7(5): 334–42.
Warszawski A, Becker R and Navon R (2007) Strategic planning for building research—a process-oriented methodology. *Journal of Construction Engineering and Management* 133(9): 710–22.
Wikman A (2006) Reliability, validity and true values in surveys. *Social Indicators Research* 78(1): 85–110.
Wikman A (2007) Context effects as an illustration of response uncertainty – a cautionary tale. *Social Indicators Research* 84(1): 27–38.
Willenbrock JH and Thomas HR (2007) History of construction, engineering, and management in the Department of Civil and Environmental Engineering at Penn State. *Journal of Construction Engineering and Management* 133(9): 644–51.

11 Making sense of 'new age data sets': researching from afar

Johan Ninan, Mathangi Krishnamurthy and Ashwin Mahalingam

Introduction

Construction project organisations operate in a complex, uncertain, and interconnected environment that restricts our ability to understand and predict their behaviour (Geraldi, 2008). As Rooke et al. (1997) note, construction processes are carried out by people engaging in concerted social action. These 'people' include managers and engineers working within projects, as well as local communities and project-affected persons acting from without (Tutt & Pink, 2019). Increasingly, construction projects are becoming more complex with the rise of 'megaprojects'. Such large projects involve complex dynamics that bring together a large number and variety of actors: engineers, consultants, owners, users, and other stakeholders. Consequently, an investigation of any phenomenon, such as innovation, safety management, competitiveness, learning process, and so on, in the built environment must involve the study of complex social systems, such as the practices that project participants use, the structures/systems that they inhabit, and/or the dynamics of the active social groups, to name a few (Chan & Raisanen, 2009). Thus, researchers who seek to understand the built environment and improve its practice must perforce study the people who inhabit projects and the interactions between them. In this chapter, we argue for expanding the study of these interactions to the realm of the digital. The construction industry is evolving to embrace digitalisation. Multiple mobile devices, such as smart phones, tablets, and other handheld devices, are prevalent in the construction industry and support communication and collaboration during the construction process (Oesterreich & Teuteberg, 2016). These devices create data every day relating to projects that are currently not explored. We make a case for the constitutive importance of new age data sets, or data compiled from digital interactions in construction research. In doing so, we also draw upon ongoing work in digital ethnography to argue for its diverse possibilities and increasing validity.

Understanding people in construction presents a unique challenge. The industry itself is a loose constellation of actors, with no defined boundaries, spanning the public and private sectors and technology service providers of a variety of ilks. To understand people in construction, researchers have used different data sources and employed diverse analysis methods from the lexicon of social science research. Arguably, the adaptation of these methods suffers from inattention to the question of the social at the heart of construction management research. As a result, far more attention is focused on a 'business-as-usual' dependence on data writ large at the cost of the complexity of retrieving data in the first place. Data are currently collected through methods such as questionnaires, interviews, or observations. Of these, interviews are the most common data collection instrument used in qualitative research, resulting in Atkinson and Silverman (1997) claiming that we live in an 'interview society'. What such methods often ignore in construction management research is the context-dependent nature of the interview and the stakes for the interviewees in participating. An aspect necessary to qualitative study, participant observation or embodied engagement (Wacquant, 2004) has been paid very little attention in the worlds of project personnel. In a sector such as construction, which is peopled at all levels, the absence of observational data to study people's behaviour is shocking and speaks of a lackadaisical approach to social science methodologies.

As noted by Pink et al. (2010), observations can provide us with insights for addressing the practical, problem-based challenges in the construction setting. Several ethnographies have indeed been reported in this domain (Oswald & Dainty, 2020), involving considerable observation of the social setting at hand, but the nature of what counts as observation or ethnography in this regard is up for debate. Traditional ethnographies involve studying stable communities over long time spans. These time spans allow researchers to observe behaviour and practices to arrive at an understanding of culture within these social settings. In contrast, construction projects are time bound and consist of a constantly changing set of actors, activities, and artefacts as projects progress from design through construction to final commissioning and operation. We therefore see a variety of innovative strategies, such as 'intense' ethnographies (Pink et al., 2010), in the construction sector that blend observational techniques with interviews and other methods of data capture to ostensibly make sense of a continually changing social setting. Nevertheless, such studies represent a minority in the compendium of research on the built environment, and as Oswald and Dainty (2020) note, suffer from a lack of reflexivity on the part of the researchers.

For construction in the modern digital era, as with several other industries, information regarding the project is created in diverse formats across different media. Social media, news articles, emails, and other digital platforms now comprise a repository of new data that can be observed to study how construction projects behave in the 21st century (Ninan, 2020), and augment our ability to conduct intense ethnographies in rapidly changing settings. In addition, digital work is a critical component of how such projects are executed today, with traditional document-based artefacts, such as blueprints and approvals, being replaced by emails, WhatsApp and other social media platforms, intranet-enabled workflows, and the like. With construction projects moving firmly into the digital age through the use of technologies, such as Building Information Modelling, augmented and virtual reality, and artificial intelligence, data compiled from digital sources, categorised as 'new age data sets' or 'new ways of seeing' (Bansal et al., 2018) can give significant insights on how construction projects are managed and how they can be managed better. Yet these are almost completely ignored, as contemporary ethnographies in construction continue to base themselves on traditional means of collecting qualitative data, such as in-person interviews.

In this chapter, we call for studies using 'new age data sets' as a necessary corrective to the current understanding of ethnographic research in construction research. We argue for new age data sets as both an additional data source as well as an alternative research paradigm to enhance our understanding of an extremely fragmented and complex field. In doing so, we show how current understandings display significant blind spots that can be partly addressed by expanding the realm of the social in project management to the realm of online communication. We also argue that the means of collecting such data is as important as the ends, and that we must pay particular attention to methods to avoid the pitfalls that watered-down ethnographic techniques have encountered in contemporary construction management research.

In the following section, we review the data sets currently employed in construction research and highlight the advantages of new age data sets. Subsequently, we discuss some of these data sets and give examples of how they can be used to understand construction better. Following this, we provide some guidance and implications for data collection and analysis, as well as some gaps in our understanding and pointers for future research.

Data sets

Different data sets are used by researchers to make sense of constructs in social science research. The main classification between these data sets are 'data got up' by the researcher and 'data that are' (Potter, 2002). Some data are manufactured by researchers through their intervention, such as in the case of questionnaires, focus groups, experiments, and interviews, and some data are found in the field by the researcher in the form of naturalistic data, such as in the case of observations (Silverman, 2013). To contrast these two forms of data, Potter (2002) suggests a 'dead social scientist test'; this involves asking: 'Would the data be the same, or be there at all, if the researcher got run over on the way to work? An interview would not take place without the researcher there to ask the questions' (p. 541). To expand on this, we look at how, for understanding the construction sector, researchers need to explore 'what people know' along with 'how they express it', with the latter involving unspoken ways of knowing (Tutt & Pink, 2019), for which naturally occurring or naturalistic data can help. However, what counts as such naturally occurring data has made rapid strides in the current environment.

Current data sources

Data from questionnaires can give us a broad idea of the field; however, they are inefficient for understanding the 'deeply held beliefs' of the sector (Green & Sergeeva, 2019). These questionnaires give very little opportunity for respondents to expand upon their answers or give unprompted opinions (Lupton, 1993). To address these deficiencies with questionnaires, the natural transition was to conduct interviews, as this provided an opportunity for respondents to build upon their answers or offer new insights, which the researcher couldn't embed in their questionnaires. However, interviews too have their inefficiencies, as questions such as 'What is significant?' or 'What do terms mean?' cannot give deep insights on meanings (Emerson et al., 2011) or unveil patterns that were previously unknown. Additionally, the question of bias is not easily handled through interviews, as the response to an interview question can change if the respondent is aware of the research interests (Silverman, 2013). To handle these inefficiencies of 'data got up', researchers moved to explore 'data that are', with the understanding that they need to focus on what people do rather than on reporting accounts of practice that respondents give (Sacks, 1992). If questionnaires help us understand 'what researchers think practitioners do', and interviews point towards 'what practitioners think they do', then observations can help unearth 'what practitioners actually do'.

New age data sets

In this digital era, we argue that 'what practitioners actually do' can best be observed by studying the online environment, given the density of digital communications in contemporary construction projects. For example, internal communications within the projects happen through emails, external communication with the project community happens through social media, and sometimes knowledge is shared through multiple online platforms active in the construction domain. To this extent, the digital becomes not just an additional avenue to explore assumptions, knowledge claims, and attitudes, but indeed an important and constitutive part of the very nature of construction projects. Researchers who observe interactions in these digital environments are therefore well placed to make sense of what practitioners actually do in the built environment. After all, the internet is a 'laboratory for the social sciences', as multiple human interactions occur in these digital habitats (Hallett & Barber, 2014). Happily for researchers, these data do not need to be obtained in real time, as is the case with traditional ethnographic observations. Most of these data are stored digitally, can be retrieved over time, and can therefore supplement any other data gathered by the researcher during the period of physical observation.

This solves another critical problem that qualitative researchers of construction processes face. Construction is an evolving process that unfolds over both time and space, as large construction sites are often geographically distributed. For instance, consider a metro rail 'project' that stretches for several kilometres interspersed with tunnels, stations, and viaducts, each with their own stories. Multiple dramatic events and 'critical incidents' (Flanagan, 1954) need to be studied and documented on such projects. Researchers therefore have to be present where the action occurs to make sense of how knowledge and practices are interwoven through 'doing', which involves people and technologies within a system of social relations (Gherardi & Nicolini, 2002). Similarly, Marshall and Bresnen (2012) note the need to understand how social practices are locally and actively constituted in construction, and highlight challenges of patterning of time and space associated with project activities. Satisfying this necessity to 'be' where the action unfolds places researchers in a delicate situation. In the aforementioned example of the metro rail, should the researchers visit a tunnel where a 'breakthrough' of the tunnel boring machine is imminent? Visit the head office where a critical tender will be opened? Or study the erection of a girder on an elevated viaduct amidst peak-hour traffic on the road below? It is difficult to predict which site is likely to be the ideal one to study a priori, and researchers often risk missing key turns of events; consequently, they are forced to rely on retrospective accounts (sometimes second-hand) to reconstruct these stories. However, in the digital era, practitioners are spending more time in the digital world, and the 'digital space' is beginning to replace the 'physical space'. Videos, WhatsApp messages, tweets, and the like can help record incidents where researchers are not present, which can then be reviewed by the researcher at a later time. In short, investigating digital data can help short-circuit difficulties related to the inability of researchers to observe multiple spatially distributed incidents at the same time.

One branch of research on digital environments deals with how these environments are creating or shifting cultures of the workplace, in other words, with the anthropology of digital media (Williams, 2018; Spitulnik, 1993). Our focus here, however, is not to study the impact of digital technologies on work practices and cultures, but to understand how such data can represent an additional arrow in the ethnographic data-gatherer's quiver as they attempt to understand social settings better.

Advantages of new age data sets

Observations obtained from the digital environment can capture high-quality data that are generally not available with other instruments. First, 'data got up' by researchers have significant participant biases in them, due either to the questions asked by the researcher, or to the inherent bias of respondents. Similarly, physical observations require the presence of the researcher or the recording instrument, which can hamper the quality of the data collected. Subjects who are aware that they are being observed modify aspects of their behaviour, as seen with the 'Hawthorne effect' (McCarney et al., 2007). Data compiled from online sources are often retrospective, and subjects are not aware of the observation, thereby removing participant bias.

Second, online data sources are digitally indexed and can inform practice longitudinally without any recollection issues with the respondents. The data collected from internet archives can be similar to longitudinal data collected through multiple interviews at different points of time (Sonenshein, 2010), thus informing us about people's changing opinions, concerns, and desires (Eysenbach & Till, 2001), which can be retrieved at any point in time.

Third, gaining access to a construction setting is difficult, and respondents treat researchers as outsiders and do not share valuable information unless researchers establish close relationships with them (Pink et al., 2010). For instance, for one of our data collection efforts, the researcher gained access to data on a rail project only when he gave friendly advice to one of the respondent's sons who wanted to apply to universities in the USA (Mahalingam, 2006).

While access in such serendipitous fashion has long been part of the romance of traditional ethnographic research, it renders the process unpredictable and overly dependent on the social worlds of the researcher. In contrast, new age data sets can be collected through data shared by the construction organisation publicly, or by getting permission from only the top management. New age data sets thus enable an exploration of new avenues that were earlier hidden to researchers due to lack of primary data.

Fourth, respondents are often inconvenienced with data collection methods such as questionnaires and interviews as they have to take time off their work. New age data sets are digital data and can be collected without causing any inconvenience to the respondents.

Finally, respondents can be fearful of sharing too much information during interviews and be at the risk of revealing some confidential data to the researcher. To mitigate this, respondents exercise extreme caution while responding to questions, and this can hamper the data collection effort. In contrast, new age data sets, such as emails, can be screened by the organisation before being handed over to researchers. While this hampers some of the truth-finding goals of research, it allows greater ease vis-à-vis ethical imperatives and helps mitigate risks to respondents. Thus, as Banyard and Hunt (2000) noted, the way in which data are collected has a significant effect on the behaviour of the respondent and the quality of the data, and new age data sets help provide unobstructive and unbiased data.

New age data sets: some examples

New age data sets can be obtained from different sources, such as social media, online news articles, emails, and other digital platforms. We discuss each of them separately below.

Data from social media

The popularity of social media, such as Facebook, Twitter, YouTube, Instagram, and other platforms, is increasing day by day. Many construction projects use social media to engage with the community by operating their own social media page. For instance, in our study of an infrastructure megaproject in India (Ninan et al., 2019), the project team used social media to routinely update the community about, for instance, the progress of the project. The following Twitter entry is an example of this: 'Track work progress as on

18-12-13. Plinth completed for a length of 17719 m on viaduct between *** [station A] and *** [station B]'. Social media can also be used for other purposes, such as marketing and branding (Sivertzen et al., 2013). For example, in our study, we saw that events such as painting competitions for school children that were conducted by the project were popularised within the local community through a series of tweets, such as: 'Painting competition at 4 pm Today!!! Topic – Go Green Metro – Timing 4 pm to 6 pm – Don't forget to bring your colors'.

Such data open up avenues to study new kinds of research questions, for example, how megaprojects strategically conduct community-based events, such as painting competitions on the theme 'go green metro' to paint their sustainability agenda, and get into the hearts and minds (Henisz, 2017) of school children. Researchers can use the natural data from social media to make sense of how a project interacts with different stakeholders via the digital platform. From this new age data, it is possible to study the multiple affordances of social media in projects, and how different stakeholders in the project setting use social media for their own agenda. In addition, such data can provide insights into organisational strategising. In our case, tweets helped us understand how project promoters framed their project to obtain support from the broader stakeholder community. Equally, tweets could also have been used to understand the perspective of stakeholders towards the project as well as the strategic and tactical interventions undertaken by the project team. Such data therefore allow the researcher to 'view' the interactions between actors in detail – project proponents and stakeholders, in this case. It is important here to remember, however, that despite arguments by many as to the fluidity of online and offline lives such that they flow from one to another (Gershon, 2010; Miller, 2013), data from social media must also be supplemented by other forms of qualitative research, such as interviews, to understand the relationship between the two.

Data from online news articles

Construction projects, particularly infrastructure megaprojects, attract a lot of media attention due to their colossal and controversial nature (Frick, 2008). Modern construction projects recognise the importance of the media in shaping the perception of the project, and have a media strategy with dedicated officers handling public relations and interactions with the media (Van Marrewijk et al., 2008). In one of the cases we studied (Ninan et al., 2020), the following quote from a project participant was not obtained through an interview, but was found as a spokesperson's quote in the media: 'Once fully operational, the total capacity of rooftops and ground mounted power systems in [metro rail organisation's] facilities will be six MW [megawatts], which will make it one of the largest on-site solar projects in India'.

As seen from the instance above, projects market their achievements through the wide reach of the news media. Here, the project highlighted its salient features and promoted these as innovative, community-centric, pioneering, and so on. Along with such positive news of the project, negative news, such as accidents, delays, corruption, and other events, can also be reported in these news mediums, and are often swiftly countered by project participants on similar fora. In the event of an accident on site, the project spokesperson reported in the media, 'The workers did wear their safety gear and other safety precautions were in place. We [metro rail organisation] are in the process of finding out how it happened'.

Analysing data available on digital media can therefore help researchers understand a project's strategic response to externalities. In addition, researchers can study news media to understand the media strategy of construction projects and how projects organise themselves towards the media. Along with the promoters of the project, project protestors also use news media to shape the narratives surrounding the project (Ninan & Sergeeva, 2021). Researchers can use this new age data set to make sense of the media strategy and the role of media in shaping the projects we see today. Usage of this data set is also crucial, keeping in mind a key aspect of the sociality of construction projects in that their reputation is constitutive of their capability to forge forward. Managing public perceptions therefore becomes paramount as one of the key aspects of project management.

Articles in the news media have long been used as a secondary data source in qualitative research to supplement direct observations and interview data. However, digital media has evolved beyond the digitisation of print media and includes blogs, online-only news outlets, and the like, leading to a potentially rich data set for researchers to access. Indeed, one can consider upgrading the status of such media data to 'primary' data sources that can be mined, as compared to their current status as 'secondary data'. After all, it is what researchers do with the data that determines whether the data are primary or secondary (Speer, 2002). This may be of particular resonance to construction projects, given the capacity of digital media to record responses, strategies, and conversations over time.

Data from emails and other forms of internal communications

Natural data for new insights on different areas of construction can also be taken from emails. During this digital age, communication within the construction project team occurs through digital mediums, often through emails. Construction projects use emails for correspondence with the multiple agencies and their employees involved in the project. In one of our studies on an infrastructure project in Australia, promotional emails were sent out to the project team, as follows:

> This week, we put the spotlight on our partnership with [name of local sports team]. 'We're proud to be partnering with the [name of local sports team]. It's an important part of our commitment to leaving a positive legacy along our project corridor and across [name of place]', says [name of person], GM, Corporate Communications. Find out how [link], through this partnership, we've been able to help more than 45 000 people in [name of place].

The project experienced intense resistance from the project community, comprised of taxpayers and people from whom land was acquired. In infrastructure projects, it is seen that the anti-project discourses of these external stakeholders percolate to the internal stakeholders, such as the project team, often thereby affecting their perception of the project (Ninan et al., 2021). To counter this, the project circulated a weekly internal newsletter via email among all employees aimed to promote the project, such as seen with the email on the partnership with the local sports team. The internal newsletters also kept track of resistance groups and their activities, and highlighted how the project, in contrast, has a wider reach: 'Great news – our [link of Facebook page] now has more page likes than the two main action groups combined [names of action groups]'.

Communication patterns have changed through the use of tools such as emails. Mailing lists and the like allow practitioners to reach large groups of people simultaneously, and communication is no longer restricted to physical presence. This then creates new networks through which information flows on projects. Hossain (2009) studied email data and demonstrated that informal network centrality confers more influence in a project than formal organisational position. Thus, data from emails can inform construction project behaviours, such as coordination, innovation, power dynamics, network centrality, and other complex constructs that were traditionally only studied through interviews and observations, and that are now subject to subtle electronic influences that cannot be noticed physically. Ramalingam and Mahalingam (2018) show how analysing emails sent over the period of a year allows us to understand how transnational design firms evolved practices to bridge cultural boundaries, for instance.

Emails are not the only internal communication tools used in projects. While project intranets have been in vogue for years, new age technologies such as WhatsApp have taken over the function of project communication and coordination. Recent research that we undertook in India (and that is yet unpublished) showed that the density of interpersonal communication was far higher in WhatsApp than in in-person meetings. While several project actors chose to stay silent in meetings, perhaps in deference to formal authority, they were extremely active on WhatsApp, providing information that was critical to the progress of the project. A social network analysis based on the WhatsApp transcript was telling, as the 'central' figures in the in-person networks were peripheral ones on WhatsApp, and vice versa. In some sense, this project exhibited dual organisational structures – one formal, as evinced in the in-person interactions, and an informal virtual structure, as evidenced in WhatsApp. The process of coordination also changed. Terse messages in slang accompanied with photographs were used to coordinate rapidly, as opposed to formal documents, presentations, or notes that were often present in review meetings.

Also important to these many routes of communication is understanding their conventions as detailed above, including, for example, the use of emojis, the extent of participation in these fora, typing cadence, and even typographical errors.

Social scientists also call for 'textual listening' (Boelstorrf et al., 2012) to be deployed with these new data sets. Such listening mimics existing understanding in fields such as cultural studies, where chosen artefacts are read as texts to be able to perceive ongoing cultural change. As mentioned earlier, chat data providing evidence of a switch in the centrality of personnel are important in understanding obstacles, catalysts, and key movers in the informal conduct of projects.

Data from other digital platforms

Data can also be captured from other digital platforms, such as innovation platforms, education platforms, and others. For instance, the i3P (Infrastructure Industry Innovation Partnership) platform allows players in the UK infrastructure sector to collaborate and share innovation stories. Data retrieved from this source can help researchers understand the most common innovations, the motivations for these innovations, different stakeholder interests in these innovations, and so on. For example, a company reported an innovation for improving safety of Heavy Goods Vehicles (HGV) in the i3P platform, as below:

> Through our work directly involved in the logistics running of the business, we can recognise the need to improve cycle safety regarding HGV vehicles. We read in newspapers daily of incidents and accidents of this type[. As] specialists in the business, we need to push these improvements through to standard fitment.

Analysing this story gives us insights on the motivation for the innovation. The employee reported that newspapers reporting incidents and accidents relating to HGV vehicles was the motivation for this particular innovation. In another instance, a company's target of a 10 per cent reduction in the carbon footprint motivated innovations, as recorded in another post on the platform: 'In order to meet the project target of an 10% reduction in [name of owner] carbon footprint, the team must continually seek innovative solutions to the daily operational needs'. Researchers can look for similar naturally occurring data in diverse platforms to make sense of innovation, safety practices, quality performance, and other issues of interest.

Table 11.1 summarises these new age data set categories and identifies the types of findings that they enable. In each of these cases, new age data sets helped researchers understand a critical social process in construction projects that they would have had considerable difficulty unearthing by just relying on interviews and direct observations.

We only discuss some of the sources of new age data sets that can be leveraged by researchers in the construction management discipline. Data from these sources provide us with new 'ores' that have to be mined to make sense of how people behave in the construction industry. Also important to remember is that these new 'ores' may provide evidence of processes, behaviours, and feelings hitherto unavailable to researchers relying on traditional methods and data sets. Additionally, as we move forward, attention to new age data sets becomes critical to understanding the future of construction management research.

I've found 'new age' data. Now what?

Barley and Kunda (2001) eloquently convinced us to 'bring work back in' as we study engineering settings. It is clear now that digital technologies are

Table 11.1 Summary and examples of new age data sets

New age data set category	Example of research using the new age data set	Unique findings enabled by the new age data set in existing research
Data from social media	Twitter and Facebook posts of a metro rail project in India (Ninan et al., 2019)	The analysis helped understand how social media is used in the construction industry for marketing, branding, and the management of external stakeholders
Data from online news articles	Quotations of construction project spokesperson reported in news articles (Ninan et al., 2020)	The news articles helped understand the project's strategic response to externalities, such as accidents and delays
Data from emails and other forms of internal communication	Emails sent over the period of a year (Ramalingam & Mahalingam, 2018) WhatsApp communications of a project in India (unpublished)	Analysing emails helped understand how transnational design firms evolved practices to bridge cultural boundaries; WhatsApp communications led to the identification of informal social networks of information exchange on projects that were very different from formal networks observed in project meetings
Data from other digital platforms	Innovation stories posted in i3P (Infrastructure Industry Innovation Partnership) platform in the UK (Ninan et al., 2022)	Different motivations for innovations were found from the innovation stories, such as reports of accidents at other sites, and an organisation's target for reducing their carbon footprint

an integral part of the work that we do (actor network theory, for instance, freely allows for artefacts such as computers to be classified as actors/actants in organisational networks), and that business happens over intranets, emails, Facebook, Twitter, and other platforms. The study of engineering organisations can therefore no longer ignore these artefacts and needs to access digital communication as a critical data source for understanding contemporary organisational phenomena. The final product of the construction process is a physical object that can be examined; however, the construction process as such can only be understood as a socially constructed phenomenon (Sutrisna & Barrett, 2007). New age data sets can give more insights on the 'social turn' in construction management research. Interpersonal relations in construction settings, such as trust, power, or coordination, can be studied through visual rituals depicted by artefacts, such as jokes, narratives, or forms of language (Gajendran et al., 2011) available in these new age data sets. Along with exploring current research areas, new age data sets inform practices that are currently not explored. For example, many conversations relating to the construction industry are only evident online and are not currently captured or analysed. Thus, communications evident in social media, online news media articles, emails, and other digital media can provide authentic and deep insights that would otherwise remain unseen. Such an exploration would help us understand how projects behave in the 21st century.

The process of accessing and analysing these data displays similarities with regard to how information is accessed and analysed in traditional qualitative research techniques. Tactically, researchers have long spoken about 'gaining entry' into a social setting and cultivating informants. Accessing new age data sets also requires strategies that help the researcher 'gain entry', as organisations will be rightfully worried about how the confidentiality of their data will be maintained. Building trust therefore remains a critical first step in the data collection process. While researchers may not need to rely on techniques, such as 'ethnographic interviewing' (Spradley, 1979), since the data are already recorded digitally and are available, there are disadvantages, particularly with regard to sifting through the data to find portions that are relevant. Thousands of emails are sent within organisations, sometimes even on a daily basis, and many may not be relevant to the researcher's question of interest. Sifting the wheat from the chaff could prove to be a time-consuming exercise. Of course, technology can come to one's aid here. Marwick (2013) suggests that hashtags can be used to bound and select relevant data on platforms such as Twitter. Henisz et al. (2013) attempt to use natural language processing as a tool to 'mine' large numbers of articles in digital media. However, none of these approaches have been perfected at the moment.

Beyond this, analysis techniques remain similar. The principles of coding that, for instance, are popular in traditional qualitative research (Strauss & Corbin, 1990) apply here as well. Emails or WhatsApp transcripts can be categorised, coded, and analysed to develop salient constructs and to understand the interplay between them. There is a tendency among researchers using these data sets to attempt quantitative analysis of their data, primarily because the volume of these data points and the ease by which they can be accessed through computational tools is high. Wang and Taylor (2018), for instance, use Twitter data to computationally model human mobility in the face of natural disasters, and Breslin et al. (2020) use Twitter to map political attention to COVID-19. Indeed, this constitutes an advantage of using new age data sets and can allow authors to better validate and ensure the replicability of their data. However, there is also the danger of taking quantitative analysis a bit too far. As Marwick (2013) suggests in her analysis of ethnographic research on and using Twitter, such 'big data number crunching' can be valuable, but often ignores aspects such as the motivation to use a particular tool. On the other hand, she notes that new age data sets can help explicate participants' 'meaning making activities' and even enable the development of 'thick descriptions' (Geertz, 1973) of interactions. However, how and why people post on these new age platforms cannot be taken for granted. The context with which a user uses these platforms is salient and cannot be ignored. As Papacharissi (2012) notes, there may be a notion of 'play' amongst tweeters where the intent of posting on this platform as opposed to engaging verbally and directly is due to the ease with which data can be discredited on Twitter. Miller (2013), in a Facebook study, suggests loneliness as a cause for actors resorting to social media. These findings indicate that the context, the nature of the individual, and their motivation need to be taken into account in the process of understanding how a social setting unfolds; therefore, a qualitative approach (perhaps in addition to a quantitative one) might reap rich benefits in understanding the implications of such data sets. In sum, the basics of data access and analysis remain similar, but the nuances of dealing with new age data sets need to be fully internalised by researchers.

The final product of the construction process is a physical object that can be physically examined; however, the construction process as such can only be understood as a socially constructed phenomenon (Sutrisna & Barrett, 2007). New age data sets can give more insights on the 'social turn' in construction management research. Interpersonal relations in construction settings, such as trust, power, or coordination, can be studied through visual rituals depicted by artefacts, such as jokes, narratives, or forms of language (Gajendran et al., 2011) available in these new age data sets. Along with a foray into new age data sets, researchers must also intensify their efforts to locate motivation in relation to data. To this extent, each set and its particularities – Facebook and identity management, Instagram and identity culture, email and corporate networking, for example – will provide rich data in new contexts for continued work.

These new age data sets can also be combined with other data, such as interviews, to give some interesting insights. For example, in earlier work (Ninan et al., 2020), we looked at news media articles to obtain data on 'what was communicated strategically to external stakeholders', and semi-structured interviews to collect data on 'what really happened at the site'. Contrasting both these data sets produced insights on the practice of strategic communication of information to external stakeholders in the case of an infrastructure megaproject. It is therefore not our contention that interviews and observations are inadequate means of data collection, nor do we suggest that new age data sets should replace these traditional means of data collection. Rather, our contention is that the construction industry is a heterogeneous constellation of actors, actions, and artefacts, and that new age data sets can complement interviews and observations with project participants, providing significant insights into on-the-job practices in a project as a primary data source and not as a secondary one. The traditional notion of such data as 'secondary' implies that it plays second fiddle to data collected during the direct presence of the researcher. Given the amount of work that happens 'out there' that is stored digitally, it is perhaps time to overturn this hierarchy and consider digital data and new age data sets as primary forms of data that cannot be ignored if we are to develop a true picture of the social settings that characterise construction projects.

It is worth pointing out here that researchers should necessarily also consider the ethical issues of using new age data sets. The practices of informed consent and privacy of respondents vary between private spaces and public spaces (Eysenbach & Till, 2001). Researchers should respect privacy and seek informed consent for private online spaces, such as those that require some form of registration or subscription and those not open to the general public. In contrast, public online spaces that are open for anyone to read may not require explicit consent, as contributors to these online fora are aware that their posts are publicly available and open to anyone to read (Dehkhoda et al., 2020). Data from these public spaces can be treated by researchers in ways similar to published material (Casselman & Heinrich, 2011). It is important to safeguard the privacy of respondents due to the traceability of quotes (Beaulieu & Estalella 2012). Here, too, there are ways to overcome traceability and make quotes 'Google-proof', such as cutting the length of verbatim quotes and hiding not only the respondent's name but also identifiers, such as place, associations, and so on. Another ethical issue with using new age data sets is the use of the published material without giving credit, thereby exploiting someone's intellectual property (Eysenbach & Till, 2001). Thus, researchers are advised to assess the ethical concerns of conducting an online naturalistic inquiry and obtain proper ethical clearance considering the risks to the respondents.

Conclusion

New age data sets comprise a new 'ore' that is currently not extracted and mined. This chapter thus calls for a research agenda in construction management using new age data sets as an important complement and a new direction to the current widespread use of interviews and questionnaires. A turn towards exploring these new age data sets can throw much light on some of the overlooked dynamics, such as identity, coordination, and strategic communication, in construction management research and reveal insights that have hitherto been hidden in conventional qualitative enquiries. Part of the reason for such data being overlooked is a traditional and conservative understanding of what counts as legitimate data from a physical and tangible social setting. In this, principles dear to digital ethnography need to lead the way such that increasing engagement among people – workers and managers – in the digital realm can no longer be treated as an additional and minor source of data. For daily life as much as for our work lives, the digital has become a constitutive arena for communication, identity formation, expression, feelings, and thoughts. Without giving this due cognizance, construction research methods run the risk of being, at best, inattentive and, at worst, downright inaccurate.

In this chapter, we describe new age data sets and ways of dealing with them in terms of collection and analysis. However, given their emerging nature, more lessons will need to be learnt and conveyed to a growing body of researchers who are aspiring to use these data. We encourage the research community in construction management to engage with these data sets and contribute to a body of knowledge on working with them in our field. Can we systematically think of new lines of enquiry that are now made possible by the presence of and our access to these data sets along the lines of those suggested in Table 11.1? Can we attempt to refine data collection and analysis methodologies, keeping both the nature of the construction industry and the data set in mind? Large projects will generate extremely large sets of data. What means can we use to make sense of this? We call upon researchers in the domain of construction management to acknowledge the primacy of these new age data sets and to work in cohesion to evolve techniques and best practices for their use.

In all cases, we argue for new age data sets as inaugurating an evolving paradigm vis-à-vis the social science turn in construction management research. Even as the social sciences grapple with, theorise, and produce guidelines for digital ethnography, as it were (Boellstorff et al., 2012; Pink et al., 2015), construction management research must necessarily delineate the implication of such in relation to ongoing work. New age data sets, in our understanding, provide fertile ground for fruitful experimentation to build a variegated toolkit for researchers in the field. They are an evolving and exciting possibility for construction research to deepen its adoption of social science research methodologies and place a sophisticated and relevant, yet pragmatic, understanding of the social at the heart of its concerns. Yet, there are risks as well. Researchers may gravitate towards 'big data' that is readily available without paying due attention to contextual factors. An overreliance on the quantum of data may lead to compromises in the methods of analysis, and consequently in the final conclusions that are drawn as researchers become intoxicated with these new age data sets. Can the construction management research community develop the skills, rigour, and stewardship to use these data sets to unlock the mysteries of a complex industry? Or will we see abuse instead, as we have seen with questionnaires and interviews? Only time will tell.

References

Atkinson, P., & Silverman, D. (1997). Kundera's immortality: The interview society and the invention of the self. *Qualitative Inquiry*, *3*(3), 304–325.

Bansal, P., Smith, W. K., & Vaara, E. (2018). New ways of seeing through qualitative research. *Academy of Management Journal*, *61*(4), 1189–1195.

Banyard, P., & Hunt, N. (2000). Reporting research: Something missing? *The Psychologist: Bulletin of the British Psychological Society*, *13*(2), 68–71.

Barley, S. R., & Kunda, G. (2001). Bringing work back in. *Organization Science*, *12*(1), 76–95.

Beaulieu, A., & Estalella, A. (2012). Rethinking research ethics for mediated settings. *Information, Communication & Society*, *15*(1), 23–42.

Boellstorff, T., Nardi, B., Pearce, C., & Taylor, T. L. (2012). *Ethnography and virtual worlds: A handbook of method*. Princeton University Press.

Breslin, S., Enggaard, T., Blok, A., Gardhus, T., & Pedersen, M. (2020). How we tweet about coronavirus and why: A computational anthropological mapping of political attention on Danish Twitter during the Covid-19 pandemic. http://somatosphere.net/forumpost/covid19-danish-twitter-computational-map/

Casselman, I., & Heinrich, M. (2011). Novel use patterns of *Salvia divinorum*: Unobtrusive observation using YouTube. *Journal of Ethnopharmacology*, *138*(3), 662–667.

Chan, P. W., & Raisanen, C. (2009). Editorial: Informality and emergence in construction. *Construction Management and Economics*, *27*(10), 907–912.

Dehkhoda, A., Owens, R. G., & Malpas, P. J. (2020). A netnographic approach: Views on assisted dying for individuals with dementia. *Qualitative Health Research*, *30*(13), 2077–2091.

Emerson, R. M., Fretz, R. I., & Shaw, L. L. (2011). *Writing ethnographic fieldnotes*. University of Chicago Press.

Eysenbach, G., & Till, J. E. (2001). Ethical issues in qualitative research on internet communities. *British Medical Journal*, *323*(7321), 1103–1105.

Flanagan, J. C. (1954). The critical incident technique. *Psychological Bulletin, 51*(4), 327–358.
Frick, K. T. (2008). The cost of the technological sublime: Daring ingenuity and the new San Francisco–Oakland bay bridge. In H. Priemus, B. Flyvbjerg, & B. van Wee (Eds.), *Decision-making on mega-projects: Cost–benefit analysis, planning and innovation* (pp. 239–262). Edward Elgar.
Gajendran, T., Brewer, G., Runeson, G., & Dainty, A. (2011). Investigating informality in construction: Philosophy, paradigm and practice. *Australasian Journal of Construction Economics and Building, 11*(2), 84–98.
Geertz, C. (1973). *The interpretation of cultures.* Basic Books.
Geraldi, J. G. (2008). The balance between order and chaos in multi-project firms: A conceptual model. *International Journal of Project Management, 26*, 348–356.
Gershon, I. (2010). *Break-up 2.0.* Cornell University Press.
Gherardi, S., & Nicolini, D. (2002). Learning the trade: A culture of safety in practice. *Organization, 9*(2), 191–223.
Green, S. D., & Sergeeva, N. (2019). Value creation in projects: Towards a narrative perspective. *International Journal of Project Management, 37*(5), 636–651.
Hallett, R. E., & Barber, K. (2014). Ethnographic research in a cyber era. *Journal of Contemporary Ethnography, 43*, 306–330.
Henisz, W. J. (2013). Preferences, structure, and influence: The engineering of consent. Global Strategy Journal, 3(4), 338–359.
Henisz, W. J. (2017). *Corporate diplomacy: Building reputations and relationships with external stakeholders.* Routledge.
Hossain, L. (2009). Effect of organisational position and network centrality on project coordination. *International Journal of Project Management, 27*, 680–689.
Lupton, D. (1993). Risk as moral danger: The social and political functions of risk discourse in public health. *International Journal of Health Services, 23*(3), 425–435.
Mahalingam, A. (2006). *Understanding and mitigating institutional costs on global projects* [Unpublished doctoral dissertation]. Stanford University.
Marshall, N., & Bresnen, M. (2012). Where's the action? Challenges of ethnographic research in construction. In S. Pink, D. Tutt, & A. Dainty (Eds.), *Ethnographic research in the construction industry* (pp. 126–142). Routledge.
Marwick, A. (2013). Ethnographic and qualitative research on Twitter. In K. Weller, A. Bruns, C. Puschmann, J. Burgess, & M. Mahrt (Eds.), *Twitter and society* (pp. 109–122). Peter Lang.
McCarney, R., Warner, J., Iliffe, S., Van Haselen, R., Griffin, M., & Fisher, P. (2007). The Hawthorne effect: A randomised, controlled trial. *BMC Medical Research Methodology, 7*(1), 1–8.
Miller, D. (2013). *Tales from Facebook.* Polity Press.
Ninan, J. (2020). Online naturalistic inquiry in project management research: Directions for Research. *Project Leadership and Society, 1*(1), 1–9.
Ninan, J., Clegg, S., & Mahalingam, A. (2019). Branding and governmentality for infrastructure megaprojects: The role of social media. *International Journal of Project Management, 37*(1), 59–72.
Ninan, J., Mahalingam, A., & Clegg, S. (2020). Power and strategies in the external stakeholder management of megaprojects: A circuitry framework. *Engineering Project Organization Journal, 9*(1), 1–20.
Ninan, J., Mahalingam, A., & Clegg, S. (2021). Asset creation team rationalities and strategic discourses: Evidences from India. *Infrastructure Asset Management, 40*(1), 1–9.

Ninan, J., & Sergeeva, N. (2021). Labyrinth of labels: Narrative constructions of promoters and protesters in megaprojects. *International Journal of Project Management, 39*(5), 496–506.
Ninan, J., Sergeeva, N., & Winch, G. (2022). Narrative shapes innovation: a study on multiple innovations in the UK construction industry. *Construction Management and Economics, 40*(11–12), 884–902.
Oesterreich, T. D., & Teuteberg, F. (2016). Understanding the implications of digitisation and automation in the context of Industry 4.0: A triangulation approach and elements of a research agenda for the construction industry. *Computers in Industry, 83*, 121–139.
Oswald, D., & Dainty, A. (2020). Ethnographic research in the construction industry: A critical review. *Journal of Construction Engineering and Management, 146*(10), 1–13.
Papacharissi, Z. (2012). Without you, I'm nothing: Performances of the self on Twitter. *International Journal of Communication, 6*, 1989–2006.
Pink, S., Horst, H., Postill, J., Hjorth, L., Lewis, T., & Tachhi, J. (2015). *Digital ethnography: Principles and practice*. SAGE.
Pink, S., Tutt, D., Dainty, A., & Gibb, A. (2010). Ethnographic methodologies for construction research: Knowing, practice and interventions. *Building Research Information, 38*(6), 647–659.
Potter, J. (2002). Two kinds of natural. *Discourse Studies, 4*(4), 539–542.
Ramalingam, S., & Mahalingam, A. (2018). Knowledge coordination in transnational engineering projects: A practice-based study. *Construction Management and Economics, 36*(4), 1–16.
Rooke, J., Seymour, D., & Crook, D. (1997). Preserving methodological consistency: A reply to Raftery, McGeorge and Walters. *Construction Management and Economics, 15*(5), 491–494.
Sacks, H. (1992). *Harvey Sacks: Lectures on conversation*. Blackwell Publishing.
Silverman, D. (2013). *A very short, fairly interesting and reasonably cheap book about qualitative research*. SAGE.
Sivertzen, A. M., Nilsen, E. R., & Olafsen, A. H. (2013). Employer branding: Employer attractiveness and the use of social media. *Journal of Product & Brand Management, 22*(7), 473–483.
Sonenshein, S. (2010). We're changing or are we? Untangling the role of progressive, regressive, and stability narratives during strategic change implementation. *Academy of Management Journal, 53*(3), 477–512.
Speer, S. A. (2002). Natural and contrived data: A sustainable distinction? *Discourse Studies, 4*(4), 511–525.
Spitulnik, D. (1993). Anthropology and mass media. *Annual Review of Anthropology, 22*, 293–315.
Spradley, J. R. (1979). *The ethnographic interview*. Harcourt Brace Jovanovich College Publishers.
Strauss, A., & Corbin, J. M. (1990). *Basics of qualitative research: grounded theory procedures and techniques*. SAGE.
Sutrisna, M., & Barrett, P. (2007) Applying rich picture diagrams to model case studies of construction projects. *Engineering, Construction and Architectural Management, 14*(2), 164–179.
Tutt, D., & Pink, S. (2019). Refiguring global construction challenges through ethnography. *Construction Management Economics, 37*(9), 475–480.

Van Marrewijk, A., Clegg, S. R., Pitsis, T. S., & Veenswijk, M. (2008). Managing public–private megaprojects: Paradoxes, complexity, and project design. *International Journal of Project Management, 26*(6), 591–600.

Wacquant, L. (2004). *Body & soul: Notebooks of an apprentice boxer*. Oxford University Press.

Wang, Y., & Taylor, J. E. (2018). Coupling sentiment and human mobility in natural disasters: A Twitter-based study of the 2014 South Napa Earthquake. *Natural Hazards, 92*, 907–925.

Williams, B. (2018). Virtual ethnography. *Oxford bibliographies*. https://www.oxfordbibliographies.com/view/document/obo-9780199766567/obo-9780199766567-0107.xml

12 Does construction need management? The case for alternative construction organization

Daniel Sage

Introduction

My purpose in this chapter is to question the assumption that management is the only way to organize construction. This is an ambitious and risky argument, particularly in a book on the future of construction management research. The careers of construction management scholars are invested in an object of knowledge called 'construction management'. Thus, it is reasonable to assume that any critical questioning of that object might be perceived as threatening to those individuals involved. This is not my intention. I am convinced that the object of knowledge called 'construction management' has never fully defined the intellectual curiosity and ambitions of those working in this field. This curiosity surfaces when scholars ask open-ended questions about whether there are ways of organizing construction that are more valuable, more sustainable, less discriminatory, and more socially responsible. In short, are there alternatives? The problem is that the only available answers to such questions seem far less radical and imaginative than the initial curiosity that inspired them. My contention is that this is because there only really seems to be one available answer: management. Project failures are to be solved by effectively managing projects to meet an 'iron triangle' of time/cost/quality targets (Ong et al. 2019). Climate change can be mitigated through optimizing construction management to reduce greenhouse gas emissions (Tang et al. 2013), and workplace discrimination can be addressed by injecting equality, diversity, and inclusion practices into construction management (Al-Bayati et al. 2017). If the problem is how to best organize construction, then the solution is almost always more and better management.

Across this chapter I question the assumption that construction has to be organized through management. My intention is to open up a space for alternatives. To be clear, I am not advocating the end of construction management scholarship. What I am proposing is that management is not the only way, or indeed necessarily the best way, of organizing construction. I think there are strong theoretical, empirical, political, and ethical arguments that support this position. Unfortunately, these arguments are seldom discussed by construction scholars. This chapter is intended to open up the field to these alternatives.

My argument is inspired by a growing body of scholarship on alternative organization (Just et al. 2021; Parker et al. 2007, 2014a; Zanoni 2020) explaining how market managerial organizing – 'permanent hierarchies of status and reward, the separation of conception from execution, the dominance of a particular form of market' (Parker 2002: 11) – has fuelled, not solved, many of the 'grand challenges' in the early 21st century: economic crises, social inequalities, and planetary ecological collapse (Klikauer 2013; Parker et al. 2014a). Rather than simply criticizing management, this body of work explores the potential (and limitations) of alternative ways of organizing, including: worker-owned cooperatives (Cheney et al. 2014); social movements (Reinecke 2018); no-fee universities (Del Fa and Vasquez 2019); intentional communities (Farias 2017); mutual insurance firms (Bousalham and Vidaillet 2018); and countless other alternative organizations (Parker et al. 2007). A handful of these studies have also explored alternative construction organizing, including, inter alia: worker-owned cooperative construction firms (Flech and Ngai 2014); temporary building by social movements (Reinecke 2018); and the preservation of buildings that formerly housed alternative organizations (Rodgers et al. 2016). While such studies offer impetus to explore alternative construction organizing, my motivation for writing this chapter also stems from the obvious yet seldom remarked empirical fact that construction does not *necessarily* require management: the organization of construction through market managerialism is a relatively recent invention associated with the rise of capitalism (Clarke 1992) and there is scarce market managerialism involved in the informal self-building that houses millions of people across the Global South (Bower 2017; Grubbauer 2018). Nevertheless, a powerful set of institutions, interests, and agendas have become concerned with capturing the organization of construction as a problem best solved by management. This capturing has inevitably limited the scope of how construction organizing is practiced as well as researched. Yet these trends are not easy to recognize, let alone challenge, because the managerialization of construction has not happened in a vacuum; rather, it is part of a long-established managerialization of society (Klikauer 2013). While a lot has been said and written to diagnose and contest the managerialization of work, organization, and society (Burham 1941/1972; Klikauer 2013; Marcuse 1964/1999; Parker 2002), the managerialization of construction has thus far been relatively overlooked (for an exception, see Sage and Vitry 2018).

One outcome of my critical exploration is that it might raise greater awareness of the multitude of ways construction might be organized, and provoke future research in and around construction management to consider the potential of these alternatives. A more radical outcome is that this chapter portends a questioning of the viability of construction management and its replacement with a new, more inclusive field of construction organization where management is one, among many, ways to organize (for a parallel argument for business and management studies, see Parker, 2018). I suspect, given the societal dominance of managerialism (Klikauer, 2013), the former outcome is more likely, but management has no divine right of ownership over the organizing activities usually labelled construction. Indeed, much construction that occurs today throughout the world, from DIY household repairs in London to the assembly of informal housing in Delhi, has little connection to knowledges and practices deemed 'construction management'. However, it can and does make a positive difference to the people and places involved. As well as tracing existing alternative construction organizing, I will also explore imagined alternative ways of organizing construction. Thus, there is also a utopian dimension to my argument. Here, utopia is not understood as a predicted, fixed future, but more as a feeling of open-ended potential to think beyond management and to not have to 'accept business as usual' (Parker et al. 2007: 301). This open-ended future refers to a recognition that we will never settle the question of how best to organize construction. Rather, we must keep asking this question. However, recognizing the utopian, open-ended scope of alternatives does not mean that alternatives do not already exist that can tangibly challenge research areas within construction management. Thus, to anchor my argument in this chapter, I will also consider the challenge that alternative organization can make to three construction management research topics: value, discrimination, and sustainability.

My chapter consists of three sections. First, I draw on wider work on alternative organization to build a case for exploring the organization of construction beyond management. Here, I explain how a particular way of organizing, termed 'market managerialism', has come to dominate construction. My argument is that this way of organizing does not just limit how construction attends to various social challenges; it intensifies those challenges. Second, I explore three alternative ways of organizing construction across three case studies: (i) Ulma Construction and Ulma Architectural Services – two cooperatives in the Ulma Group of the Mondragón Corporation; (ii) The Nant-y-Cwm Rudolf Steiner School in Pembrokeshire, Wales (UK); and (iii) Colin Ward's anarchist approach to construction. Across each of these cases I examine how market managerialism is challenged and what construction management scholars might take from these alternative ways of organizing. By way of conclusion, I draw on these cases, and the wider literature on alternative organization, to propose an agenda for researching alternative construction organizing.

The case for alternative construction organization

Alternatives to what and why?

The two most important questions to address at the outset of any discussion of alternative organization are 'alternatives to what and why?'. The simplest answer to the first question is that alternatives are counterpoised to the ideological combination of managerialism and capitalism (Parker et al. 2014a; Zanoni 2020). Parker (2002) defines this confluence 'market managerialism' and elaborates it with three meanings of the word 'management': (1) an occupational group involved in the co-ordination of people and things who receive higher status and reward; (2) a process of controlling people and things into an order that instrumentally separates the actual work being undertaken and the activity of control to maximize profit; and (3) academic knowledge that legitimizes managerial status and control by appealing to the universal inevitability of an idealized 'market' – populated by self-interested *homo economicus* labouring for the best wage to fulfil insatiable consumer desires. Alternatives are needed because, despite increasing the material standard of living for many, market managerialism renders durable a destructive politics of unlimited economic growth (Parker et al. 2014a). This is said to generate: individualization, alienation, anomie, and psychological disorders for workers and consumers (including managers); planetary ecological degradation; the marginalization of democracy by corporate and governmental interests and agendas; the expropriation of commons (e.g. air, water, forests, genetic code); systemic economic crises (as social and ecological systems are devastated); and enormous wealth inequalities (Klikauer 2013; Parker et al. 2014b; Zanoni 2020) – such that 2153 billionaires now own more wealth than 4.6 billion people, or 60 per cent of the world's population (Oxfam 2021).

Before elaborating on what forms of alternative organization are viable and valuable, it is salient to reflect on the extent to which market managerialism is evident in construction management practice and research. Parker's (2002) three-fold definition of market managerialism (as occupational group, technology of control, and form of knowledge) is useful here. First, across construction there clearly exists a distinct occupational group of 'managers' (employed under job titles such as 'site manager', 'project manager', 'construction director') who are tasked with co-ordinating plant equipment, finance, sub-contractors, suppliers, clients, building inspectors, materials, and so on. And this group almost always receives higher status and financial rewards than those directly undertaking construction tasks. According to the Royal Institution of Chartered Surveyors, in 2018 the average salary in UK construction management was over £80 000 (Construction Enquirer 2018), which is around double the average salary across construction (Linear Recruitment 2019). Second, managers are formally responsible for controlling construction to generate profit. Almost all construction firms exist to enable management to maximize wealth for shareholders or private owners. Thus, advancing Corporate Social Responsibility (CSR) in construction depends not on winning moral arguments but on building a business case for how CSR will increase profits (Loosemore and Heng 2017). Although construction perhaps remains less suspectable to formal managerial control than some other industries (Styhre 2011), informal managerial controls are equally capable of serving profitmaking (Löwstedt and Sandberg 2020; Sage et al. 2019).

Third, much construction management research legitimizes managerial status and control by appealing to the universal inevitability of 'the market' and profit-seeking 'firms' and 'clients'. Thus, concepts of value in construction management research revolve around the inevitability of the market and management as principles of construction organization. Tools such as 'Earned Value Management' (Ong et al. 2019) and 'linear planning' (Russell et al. 2014) are developed to help managers maximize profits by monitoring and predicting the financial returns on construction decisions and strategies. Diversity management approaches are proposed to equip managers with understandings of national cultural differences so they can develop more effective health and safety policies and thus more predictable (i.e. profitable) project outcomes (Al-Bayati et al. 2017). Other strands of research suggest how overcoming discrimination can improve business profitability (e.g. Francis 2017; Sang and Powell 2012). And sustainable construction researchers develop managerial tools, such as predictive simulations (Tang et al. 2013) and conceptual frameworks (Hill and Bowen 1997), that promise a way of selecting decisions and strategies that increase positive ecological impacts in ways that are cost neutral or increase profits (e.g. Tan et al. 2015; Duong et al. 2021). In short, market managerialism pervades construction management research. But is this a problem? And if it is a problem, then what is to be done? Such questions are relative. There are no universal measures of the value of a mode of organizing. If the aim of construction management research is to advance concentrations of status, power, and wealth, then market managerialism is successful.

However, the pursuit of these values has created a world with yawning social inequalities, endemic health and social problems, and the threat of planetary ecological collapse (Klikauer 2013; Parker et al. 2014b; Zanoni 2020). This is why I think it is at least worth exploring alternatives. The question then becomes, how can we judge a good alternative from a bad one?

Evaluating alternatives

The challenge is that many alternative ways of construction organizing are undoubtedly worse than market managerialism – such as the construction of Jonestown by members of the People's Temple cult in the late 1970s, where 909 people committed mass suicide, or the use of forced Jewish labour in highway construction in Nazi Germany. As Parker et al. (2014c: 629) explain, many alternatives to market managerialism – slavery, terrorism, war, riots, fascism, feudalism, and death cults – are clearly not preferable. Thus, Parker et al. (2014c) propose three principles to judge alternative organization: autonomy, solidarity, and responsibility. The first principle, autonomy, is defined as a notion that 'individuals should have choices about some of the most important ways in which they live their lives' (Parker et al. 2014c: 629); this principle is intended as an antidote to totalitarian organizing through rules and coercion in Nazi Germany and Jonestown. The second principle of solidarity counterbalances the first and corresponds to collective duties and commitments to others – recognizing that the individual freedom promised by libertarians is meaningless, except as part of a collective. That is, positive 'freedoms to' (e.g. *to* speak/protest/self-determination) cannot be achieved if society does not safeguard negative 'freedoms from' (e.g. *from* persecution/violence/discrimination; Parker et al. 2014c). The third principle is responsibility. This refers to the responsibility we have 'to the future – to the conditions for our individual and collective flourishing' (Parker et al. 2014c: 632). This principle is most obviously related to concerns around ecological sustainability, but it might also relate to concerns around avoiding future political tyrannies. Alternative organizing involves sustaining all three principles as open-ended questions to be always asked and never finally settled because 'Too much concern for ourselves ends up as possessive individualism and selfishness; too much direction from others and bending to the collective will is a form of coercion; and too many promises about the ideal future neglects the mucky problems of the present' (Parker et al. 2014c: 633).

In exploring these questions, we might encourage radical alternatives outside/beyond capitalism (Zanoni 2020) or more evolutionary micro-emancipations within market managerialist organizing (Parker and Parker 2017). Neither strategy is without risk, whether antagonistic disengagement or capitalistic co-option (Just et al. 2021). Given that it is probably 'easier to imagine the end of the world than the end of capitalism' (Jameson 2003), I suggest that both strategies are likely required. In the next section I explore the value of alternative construction organizing by discussing three cases. In selecting my cases, my assumption is that construction organizing at all scales and speeds, whether the small and fast assembly of a brick wall or the large and slow organizing of life in a city, is involved in how the politics of market managerialism are 'made durable' (Parker et al. 2014c: 634). Thus, my cases on alternative construction organizing are selected to encompass all speeds and scales of construction organizing.

Exploring alternative construction organization

It is beyond the scope of this chapter to provide a comprehensive survey of alternative ways of organizing construction. Instead, I will explore three alternative organization case studies. These cases were chosen to reflect different geographical scales, theoretical influences, life activities (schooling, dwelling, working), and forms of engagement/disengagement with market managerialism. These cases are all informed by existing social science and humanities research that has mostly been overlooked in construction management scholarship.

Construction cooperatives in the Ulma Group

The Ulma Group is an industrial division of the Mondragón Corporation: a global federation of worker cooperatives that originated in 1956 in the town of Mondragón in the Basque region of Spain. The Mondragón Corporation is the largest worker-owned cooperative group in the world, employing 80 000 people (65 000 in Spain and 15 000 in another 150 countries) and generating sales of € 11.4 billion (Mondragón 2021a).

The Ulma Group has been part of the industrial division of Mondragón since 1961 and now employs over 5000 people in manufacturing, construction, and architectural services with revenues of over €700 million. Annual profits in the industrial division of Mondragón total over €230 million. The Mondragón Corporation and Ulma Group are owned by partners who are also workers. Partners are represented within executive decision-making bodies, such as the Cooperative Congress (Mondragón Corporation) and the General Assembly (Ulma Group). These bodies select senior management teams and also make strategic decisions around pay differentials, mission, and values, and they set and review management performance targets. The core values of the Mondragón Corporation are democracy, solidarity, participation, and inter-cooperation. These values mean that one worker has one vote in partner decision-making bodies (democracy), that pay differentials between workers and management are no more than 1:6 (solidarity), worker-owners are involved in all strategic decisions (participation), and funds/resources are shared between businesses within Mondragón to protect jobs during economic downturns (inter-cooperation; Mondragón 2021a). The Mondragón Corporation and its cooperatives also invest significantly in community and welfare programmes. For example, the group currently owns and runs several schools, technical colleagues, a university, 14 research and development centres, and a social security system offering enhanced pensions, health care, and social security for its employees. Mondragón also contributes around €26 million to wider social programmes, from global humanitarian charities to solar farms in the Basque region, including through the Ulma Foundation, which is funded by the Ulma Group (Mondragón 2021b).

The Ulma Group itself contains nine business units, with two of these units – Ulma Construction (UC) and Ulma Architectural Services (UAS) – operating in the construction sector. Both of these units supply specialist products and services across the world, with UC focusing on advanced formwork and scaffolding, and UAS on manufacturing and designing prefabricated drainage systems, architectural precast, external wall systems, and ventilated facades. By supplying specialist products that require skilled supporting services, these units support higher-skilled employment across offices, factories, global logistics centres, and on-site support teams. UC employs over 2000 people and operates in 50 countries, while UAS employs 180 people across four countries. Both units also work closely with Mondragón-owned research and development sites, schools, and the Mondragón University to develop new products, services, and highly skilled workers.

Despite their similarities, there are notable differences in how the two units publicly identify with cooperativism. UC does not mention cooperativism at all within its marketing materials. Instead, emphasis is given to valuing customer 'satisfaction', 'needs', 'efficiency', 'deadlines', and 'profitability', as well as the 'value', 'safety', 'quality', 'simplicity', and 'reliability' of their products and services (UC 2021). By contrast, UAS explicitly promotes cooperative organization as part of their corporate identity. As their online corporate video explains: 'Ulma Group is part Mondragón Corporation – the largest cooperative group in the world – a people-based project in which we have common values and clear commitments to innovation, continuous improvement and sustainability' (UAS 2021). UAS marketing also frequently blends cooperativism and capitalism, as evidenced in their mission 'to develop a shared cooperative project that is profitable over time, seeking maximum customer satisfaction, to generate employment and wealth in the environment' (UAS 2021).

When viewed through Parker et al.'s (2014c) criteria for alternative organizing as autonomy, solidarity, and responsibility, both units constitute an alternative to market managerialism. Both UAS and UC support a higher degree of worker autonomy than market managerial organizations (i.e. the Mondragón principles of participation and democracy that enable workers to influence strategic decision-making). Solidarity is similarly supported through low pay differentials between managers and workers and through the sharing of funds/resources across other Mondragón units during economic downturns. And responsibility for future generations is enabled through a broad range of educational, environment, and social programmes.

Despite these achievements, UC and UAS face challenges in sustaining their cooperative values. Significantly, UC's rapid international expansion from the 1990s was not accompanied by a global roll out of its cooperative organizational model. In keeping with most units in the Mondragón Corporation, worker-owners in UC decided not to export their cooperative business model. Flech and Ngai (2014: 671–3) cite multiple explanations for this decision, including the need to sustain flexibility and control over (riskier) global business investments (i.e. to enable new businesses to be more easily closed or sold), as well as inconsistent global legal provision and cultural support for cooperatives. In other words, market managerial organizing was instrumentalized globally to protect the autonomy, solidarity, and responsibility of cooperativism among Basque worker-owners. To offset such contradictions and a degradation of cooperativism, there have been recent efforts in UC to increase worker participation in non-cooperative international subsidiaries. These attempts include mandating worker representation in strategic decisions, requirements for information-sharing with workers, strong CSR commitments, and investment in staff benefits (Flech and Ngai 2014). Although these measures are nascent and limited, they do constitute an attempt to promote cooperative values if not organizational structures.

Evidence from UAS suggests that the degradation of cooperative organizing can also occur closer to home. In their quantitative study of workers at UAS in the Basque region, Uzuriaga et al. (2018) found significant differences in perceptions of cooperative beliefs and behaviour in UAS, with factory workers expressing negative experiences and office workers more positive. Uzuriaga et al. (2018) explain these differences as stemming from a relative lack of task autonomy, poor education, and weaker career progression options among factory workers. This finding suggests that the high-level cooperative principles and values of Mondragón cannot inoculate workers against the alienation, anomie, and psychological disorders that are said to define capitalist work (Parker et al. 2014b). The problem Mondragón faces, as Heras-Saizarbitoria (2014) explains, is that high-level cooperative principles and values are too often locally instrumentalized by managers to promote an image to gain socially aware customers, or by employees to individualistically safeguard their employment. The solution Heras-Saizarbitoria (2014) proposes is for managers and workers to recognize that 'those principles and values are more than a means to an end; they are an end in themselves' (660). One way to foster such an ethos might be to go beyond creating alternative organizational forms and structures, as with UC and UAC, and develop alternative ways of organizing individual acts of construction that resist means–ends thinking. Such possibilities are explored in my next case.

The Nant-y-Cwm Steiner School

The Nant-y-Cwm Steiner School was founded in Pembrokeshire, West Wales, in 1979 by a group of parents who supported the education philosophy of the Austrian social reformer Rudolf Steiner. The school forms part of a global network of 1100 Steiner Schools, of which 25 are located in the UK. Steiner schooling aims 'to provide an unhurried and creative learning environment where children can find the joy in learning and experience the richness of childhood rather than early specialisation or academic hot-housing' (Steiner 2021). By encouraging child-centred, kinaesthetic (artistic and movement) learning, Steiner schooling rejects market managerial approaches to education where schools compete to achieve league table rankings through high-stakes testing and an intensified and cost-controlled labour process (Stevenson and Wood 2013). Non-managerial thinking also extends to the way Steiner schools are run by a team of experienced teachers and parents rather than single head teachers.

In exploring the Nant-y-Cwm School as an alternative construction organization, I draw on the work of the cultural geographer Peter Kraftl. Across two papers, Kraftl (2006a, 2006b) explains how the school was conceived by a parent architect while other parents volunteered their labour, first converting an old schoolhouse in 1978 and then building a new kindergarten in 1990. Consistent with Steiner's philosophy, locally sourced materials and vernacular styles were also used to create a structure sensitive to the local ecological and cultural environment (Kraftl 2006b: 934). The ethos of Steiner schooling also led the architect to reject the managerial assumptions that grip contemporary construction:

> [Building the school] involved the use of very loose architectural sketches, merely used to stimulate ideas amongst the whole group of volunteer parents. In practice, the kindergarten building evolved from its ideal design ... as bricks were laid, and plaster was applied, such that it was imbued with a kind of contingent, sculptural, performative dynamism – part-deliberately reflecting the school's curriculum. (Kraftl 2006a: 496)

Kraftl (2006a, 2006b) explains how the process of building actively performed the ethos of school: parents crouched down to become children to 'measure' and feel out the design of protective learning spaces; the parents then enacted their own kinaesthetic learning as they creatively moulded shapes from the render to construct those spaces; children then later connected with notions of community as they touched a door handle or wall their parents made years earlier. Undertaken in this way, construction organizing performs a womb-like learning organization that is protective, gentle, and nurturing (Kraftl 2006a) – mirroring the first stage of Steiner schooling. The division between the organization of construction and the construction of an (alternative) organization of learning is dissolved. Ends become means, and means become ends. Construction is organized but far from managed.

The Nant-y-Cwm Steiner School offers a glimpse of an alternative way of organizing construction that encompasses many of the principles of autonomy, solidarity, and responsibility set out by Parker et al. (2014c). The parent architect encouraged parent labourers' *autonomy* over their work in a way seldom seen on most building sites; parent labourers enacted *solidarity* with each other as they built a community with shared values; and the *responsibility* of all parents to their children's futures and the locality was performed through the construction process (Kraftl 2006a, 2006b). However, although Nant-y-Cwm School presents a strong alternative to the managerialism that pervades much education, it is less obviously post-/anti-capitalist. This is because almost all UK Steiner schooling is dependent on the contribution of fees from parents to remain independent of the market managerialism of state education. Financial expediency imbues accommodation with capitalism – potentially reproducing social inequalities as Steiner schooling is commoditized by middle-class, socially liberal parents (Kraftl 2006a, 2006b). What is more, Kraftl (2006b: 940) explains how the strong sense of solidarity forged in the initial construction of the school can also alienate new families, thus risking the very financial resources that enabled that solidarity. Despite these challenges, the Nant-y-Cwm School demonstrates how alternative construction organizing can help generate radical alternative ways of learning and living in a way that goes beyond the more incremental alternatives found in the Mondragón Corporation. But could such cases really be scaled up, and what might that accomplish? In my final case I will explore these vital questions with the anarchist thinker and trained architect, Colin Ward.

Colin Ward's anarchist approach to construction

Colin Ward's (1924–2010) interest in building began with his training under the arts and crafts architect Sidney Caulfield in the 1930s, and then while working for the modernist architect Peter Shepheard in the 1940s and '50s (Guardian 2010). In his later life, Ward would take on the role as an educational officer in the Town and Country Planning Association – a charity dedicated to improving planning to serve the needs of people. His parallel interest in anarchism began during his wartime posting to Scotland, where he befriended anarchist thinkers involved in the anti-war newspaper *War Commentary* and the anarchist publisher Freedom Press. During his architectural career, he developed his anarchist work with Freedom Press, including during his time as Editor to their newspaper *Freedom*, and by publishing 31 books on architecture, self-build housing, education, children's play, town planning and water distribution, as well as anarchist theory (Guardian 2010).

Across his career, Ward argued that society can only be organized in the interest of people if it is organized without permanent top-down authority, whether governmental or corporate (Ward 1973). However, Ward, like all real anarchists, was no proponent of chaos or disorder. Indeed, as he explains, anarchism is not the opposite of organization, it is the absence of authority (Ward 1966, 1973). Ward proposes that there are two kinds of organization: 'There is the kind which is forced on you, the kind which is run from above, and there is the kind which is run from below, which can't *force* you to do anything, and which you are free to join or free to leave alone' (Ward 1966: 1, emphasis in original). His argument is that spontaneous, temporary, and local organizing is both closer to human nature and better equipped to serve the needs of people and avoid the suffering already caused, and yet to be caused, by the corruption, ineffectiveness, and violence of authority. Ward's anarchism is not presented as a speculative utopia, but rather as a possibility within existing society – 'a seed beneath the snow buried under the weight of the state and its bureaucracy, capitalism and its waste, privilege and its injustices, nationalism and its suicidal loyalties, religious differences and their superstitious separatism' (Ward 1973: 18). For Ward, authority in all its forms (socialist state, neoliberal state, capitalist corporation, public bureaucracy) fuels social inequalities, environmental damage, war and violence, and worker alienation. For Ward (1973), 'Our task is not to gain power, but to erode it, to drain it away

... we have to build networks instead of pyramids' (26). Such anarchism does not prohibit the emergence of leaders, but a leader of a group should only be temporarily appointed by that group to tackle a specific issue, and not imposed externally as a permanent authority with higher status and reward (Ward 1966). Similarly, although Ward suggests that localism is vital to anarchism, autonomous groups can decide to federate and engage in mutual aid across wider scales, including to tackle poverty and planetary ecological concerns (Ward 1973, 1996).

Ward's anarchism, with its emphasis on autonomous decision-making, solidarity through mutual aid, and ecological responsibility to future generations, clearly chimes with Parker et al.'s (2014c) criteria for alternative organizing, but how does it inform alternative construction organizing? Drawing on his architectural background, in some of his early writing, Ward (1966, 1973) explicitly problematized managerialism within architectural practices. For example, he cites a Royal Institute of British Architects (RIBA) survey that finds that autocratic approaches to architectural design – orientated around a lead architect working to a client brief – are less successful than dispersed (quasi-anarchist) approaches where a leaderless group of architects exchange ideas in a free-flowing manner and focus on the needs of building users (Ward 1966). What the RIBA survey shows is that, although the latter approach may be less profitable for the architects involved, it invariably results in a building that is quicker, cheaper, and easier to design, construct, and maintain. This decentralized process also increases job satisfaction and reduces staff turnover within architectural practice (Ward 1966, 1973).

In his later work, Ward's commitment to anarchism led him to deeper provocations that challenge the division of labour across the construction sector, especially the separation of work between architectural practices, construction contractors, clients, and end-users (Ward 1996). These divisions were challenged because they all buttress authority: 'one group of people makes decisions, exercises control, limits choices, while the great majority have to accept these decisions, submit to this control and act within the limits of these externally imposed choices' (Ward 1966: 67). To explain his critique, Ward frequently refers readers to the thoughts on housing of the anarchist architect John Turner:

> When dwellers control the major decisions and are free to make their own contribution to the design, construction or management of their housing, both the process, and the environment produced, stimulate individuals and social well-being. When people have no control, nor responsibility for, key decisions in the housing process, on the other hand, dwelling environments may instead become a barrier to personal fulfilment and a burden on the economy. (Turner, cited in Ward 1996: 101)

Notwithstanding notable examples such as Nant-y-Cwm School, it remains the case that most building occupiers (e.g. home owners, factory workers, office workers) have very little influence over the design and construction of the buildings they spend their lives dwelling within. Ward (1973, 1976) argues that this is particularly the case for housing in richer nations where, despite massive investment in housing, most people in society live in sub-standard buildings that fail to serve their needs, while many of the poorest in society remain homeless. This is said to be because housing the poor or prioritizing all user needs is not profitable to capitalists, and state housing solutions are too 'top-down' and bureaucratic to register the complex needs of people (Ward 1976: 9). By contrast, in 'poorer nations', informal self-built housing is regularly erected on seized land to house many millions of people (Ward 1976). While often derided as disorganized, crime-ridden, and chaotic, Ward (1973, 1976) cites counterevidence that these self-built places promote levels of autonomy and solidarity that are often inaccessible in 'richer nations' where authorities control land ownership and building regulations. Such informal construction facilitates individual autonomy as poor families work together to construct houses that evolve with their financial, social, and psychological needs and opportunities; this spontaneous local solidarity also extends to creating social and education programmes, including informal 'police' forces and schools (Ward 1973, 1976). Contemporary research continues to support positive assessments of informal construction in the Global South (Bower 2017; Grubbauer 2018).

In developing his thesis, Ward also explores lesser-known examples of self-building in the UK, including the notable case of the illegal adaptation of British military camps by homeless families in need of housing after the Second World War (Ward 1973, 1976). In 1946, over 40000 homeless families, including many war veterans, invaded empty British military camps and assembled new residences by adapting military buildings. The Labour government initially criticized these families and then decided to support them, even instructing construction operatives to adapt the military camps to house more families. Ward (1973, 1976) explains how the initial self-built communities in the camps possessed higher levels of autonomy, solidarity, and well-being than those later housed with government support. Another of Ward's instructive cases is the 'plotlands' of South-East England (Ward 1976, 2004). These coastal communities of small, self-built bungalows were constructed between 1870 and 1939 by working-class families from East London as holiday homes. During the Second World War they offered shelter from bombing raids, and many families became permanent residents up to the 1980s. The ability of these families to autonomously adapt their houses as their needs and means evolved was complemented with a strong solidarity between 'plotlanders' to provide shared services (Ward 2004). Such cases offer Ward (2004) evidence that anarchism could provide an 'Arcadia For All', even in richer nations (Bower 2017).

Ward's anarchism aligns with the other two cases by demonstrating how alternative ways of organizing can help us rethink the assumption that market managerialism is the only, or indeed best, way to organize construction. Each of these cases advances values of solidarity, autonomy, and responsibility in ways that challenge modes of organizing orientated towards managerial hierarchies and unrestricted profitmaking. What is more, these cases are not eccentric experiments. The industrial democracy of the Ulma Group is prospering within capitalist global markets. The Nant-y-Cwm Steiner School has been open for over four decades, delivering education to a small community and employing skilled professionals. And plotland settlements, such as Studd Hill in Kent, continue to exist as models of successful self-built communities in the UK (Bower 2017).

Researching alternative construction organization

All the cases discussed above offer evidence that tangible and positive differences can be made to people and places by organizing construction in ways that, however modestly, resist the market managerialist mantra, 'There Is No Alternative' (Parker et al. 2014c: 625). By way of conclusion, I will draw on these cases and theories of alternative organization to elaborate a research agenda for alternative construction organizing. This discussion is structured into two parts. First, I discuss broader challenges and opportunities in researching alternative construction organization. Second, I revisit the research topics of value, discrimination, and sustainability to explain how alternative construction organization can contribute to current construction management research.

Challenges and opportunities

One of the most important challenges in developing research on alternative construction organization concerns disciplinary divides. As I have discussed elsewhere (Sage and Vitry 2018), construction has been intellectually debased across Western thought since at least Plato's image of the 'master builder' opposed the 'philosopher' to define a respective pragmatic 'directive' and nobler 'calculative' knowledge. Over the centuries, this split informed a Cartesian separation between architectural and construction professions, eventually fuelling the managerial framing of construction knowledge as distinct from the theoretical development of the social sciences and humanities (notwithstanding the recent interest in instrumentally applying social science and humanities theories to tackle industry problems). Likewise, with some notable exceptions (summarized in Sage and Vitry 2018), this division has fuelled a relative lack of interest by social science and humanities scholars in construction (as opposed to architecture) as an object of analysis. This situation poses challenges for any construction management researcher interested in alternative construction organizing, as there have been few well-researched cases of alternative construction organizing. But this challenge also generates an opportunity wherein construction management researchers might develop such studies to contribute to wider debates around how alternative construction organizing can generate alternative social organizing.

However, in developing such research, there is also a risk that only more modest alternatives, such as the industrial cooperatives of the Ulma Group, might be deemed admissible within construction management scholarship, and then only if they are benchmarked against the performance indicators of more mainstream market managerial approaches. Such comparative work is of course important to build a case for alternatives, but I hope it does not delimit the horizons of opportunity for alternative construction organizing research. Construction management researchers could also add significant value in exploring more radical alternatives. For example, they might explore whether anarchist self-building is no more than a 'historical curiosity' (Bower 2017) by taking seriously such alternative ways of organizing construction and exploring how they compare to, and perhaps interact with, market managerialism. Grubbauer's (2018) study of the interaction between corporate building supply retailers and informal self-built housing in Mexico City offers one such example. Such research can help empirically evaluate Ward's assumptions that anarchic forms of decentralized construction organization can better serve the needs of users and society at large, as well as being cheaper, faster, and more sustainable than market managerialist approaches. Such research may also help 'decolonize' construction management research by challenging neo-colonial assumptions that Western scholars can only learn *about* not *from* the Global South (Frenkel 2008). More widely, this research might reveal how the organization of construction is far more implicated in sustaining political ideologies and social relations than usually considered, and thus how reorganizing construction can reorganize society (Sage and Vitry 2018).

Another challenge for alternative construction organization researchers concerns the need to understand the contexts in which such alternative construction organizations might prosper. Recent research on alternative organization has developed beyond recording isolated case studies of alternatives (e.g. Parker et al. 2007) to understand and actively nurture what might be termed the 'conditions of success' for alternative organization. Thus far, this research has examined how: inter-organizational collaboration can enable cooperativism (Bousalham and Vidaillet 2018); intermediary organizations shape social entrepreneurship (Dey et al. 2016); money can be re-socialized through egalitarian and interpersonal models of exchange (Farias 2017); and academics can support allies in alternative organizations (Parker and Parker 2017). Such contextual research is vital because merely describing alternatives in a vacuum will not help bring them into being. Rather, what is required is an understanding of how their positive impacts can be realized within wider socio-economic contexts.

Perhaps the most obvious challenge for construction management scholars is that this work is explicitly political in a way that most construction management research appears not to be. Or, as Parker and Parker (2017) explain, researching alternative organization involves the 'clear taking of sides in a struggle' (1383). But then surely all academic research is political? Are we not always 'accountable for the theorizations we produce, the research areas, questions, designs and methods we choose, the content and mode in which we transpose and disseminate research in our teaching and in consulting, media appearances and community outreach' (Contu 2019: 1457)? This is true, but it is difficult to deny the weight of the 'snow' on the 'seed', as it seems entirely possible to undertake construction management research, assume market managerialism as the only way to organize construction, and then reject any reading of this work as being political. The problem is that the politics of market managerialism (hierarchies of status and reward, logics of authority, images of self-interested *homo economicus*, etc.) have been rendered invisible because there seems to be no alternative. I am also sure that, for some scholars, admitting their political position is pro-market, pro-management, pro-capitalist is not a problem at all. Indeed, in making the case for alternatives, I fully acknowledge that, under certain conditions, set against certain alternatives, market managerialism might be the preferable way to organize construction. It is also true that market managerialism has uplifted the living standards of many people (Parker et al. 2014b). However, there is a political imperative at work in my chapter because there is growing evidence that market managerialism is not only of limited help in facing early-21st-century 'grand challenges' (e.g. economic crises, social inequalities, and planetary ecological collapse), but is deepening those challenges (Klikauer 2013; Parker et al. 2014b; Zanoni 2020). Given, as Ward (1996) once put it, that 'there are, and always have been, alternatives' (11), what I am proposing is to avoid the closure of construction organization around market managerialism. To advance this utopian potential, I will conclude by discussing how research on alternative construction organization can contribute to research around value, discrimination, and sustainability.

Towards a research agenda for alternative construction organizing

Research on values in construction management has increasingly recognized how practitioners must negotiate and pursue a range of social, ecological, and cultural values beyond the 'iron triangle' of project cost/time/quality requirements (e.g. Berg et al. 2021; Kuitert et al. 2019). However, typically these values, such as aesthetics, buildability, and sustainability, remain subservient to the 'bottom line'. As Leiringer et al. (2009) explain, 'Aspirations of providing through-life value for clients must be balanced against responsibilities to provide returns to shareholders. And in truth, the former is only worth doing if it contributes to the latter. Otherwise investors will take their capital elsewhere. This is the nature of the marketplace' (283). This situation means that, while managers may express values such as aesthetics, sustainability, and buildability (Berg et al. 2021), these values are always already circumscribed within a market-orientated logic. This does not always mean that non-financial values cannot be pursued, but it does mean they can only be pursued insofar as they either increase or do not impede profits (Leiringer et al. 2009). My contention is not that alternative construction organizations, such as the cooperatives of the Ulma Group or groups of informal housing self-builders in the Global South, do not experience financial limits. They too must operate within resource constraints. However, within alternative organizations, financial values can be more readily instrumentalized to serve non-financial values. Alternative values – democracy, solidarity, participation, and inter-cooperation (Mondragón 2021a) or 'mutual respect' and 'mutual aid' (Ward 1973: 119) – can have more authority than market managerial values of authority, competition, status, and profit maximization. The construction units of the Ulma Group provide evidence that this shift in values is tangible. Pay differentials are reduced, enhanced social security benefits can be offered to employees, investment is made in multi-million-euro local sustainability schemes, firms in different sectors are protected during recessions, and funding is provided for an entire vocational education system containing schools and a university. Cooperatives, such as UC, may still market themselves to clients around a familiar set of market managerial values of customer 'satisfaction', 'needs', 'efficiency', 'deadlines', and 'profitability', yet they do so to advance a set of non-financial values as part of a cooperative collective. The instrumentalization of financial values to serve non-financial values is only possible because these organizations are premised on ideological values of autonomy, solidarity, and responsibility that challenge market managerialist ideologies. If construction management scholars are interested in advancing a more holistic values agenda, such cases are surely worthy of more attention.

Another topic in which research in alternative construction organization can generate significant contributions is workplace discrimination. Despite the promotion of the business case for equality, diversity, and inclusion (Francis 2017; Sang and Powell 2012), there has been consistent evidence within construction management that market managerialism fuels discrimination. In their study of women's construction careers, Dainty et al. (2000) explain how, 'through their progression and informal interactions with senior management, men had the potential to increase their organizational power, with which they could maintain women's under-achievement' (248). Research also shows how male managers exploit their more senior management roles to bully and sexually harass female colleagues (Watts 2007; Wright 2013). Unsurprisingly, given these associations between construction management and discrimination, Bridges et al. (2020) recently called for equality interventions that are '"bottom up" (mentoring, networking, alternative management structures, supportive policies)' (911). The difficulty is that such practices, while well intended, may not be enough. Perhaps market managerialism, and its hierarchies of reward, status, and power, inescapably constrains initiatives to tackle discrimination. Unfortunately, existing research on alternative construction organization (e.g. Kraftl 2006a, 2006b; Flech and Ngai 2014; Reinecke 2018; Uzuriaga et al. 2018; Ward 1973, 1976) does not explore practices of discrimination. Thus, construction management researchers might turn towards a wider literature that has explored the potential (and limitations) of alternative organization to challenge workplace discrimination (e.g. Priola et al. 2014; Vitry 2020). Given the persistence of workplace discrimination (Hasan et al. 2021) and market managerialism, such alternatives should be taken seriously.

Research on alternative construction organization also offers a strong contribution to debates around sustainable construction (Duong et al. 2021; Hill and Bowen 1997; Jones et al. 2010; Tan et al. 2015; Tang et al. 2013). Almost all sustainable construction researchers seem to accept the notion that sustainability should take for granted management hierarchies and not reduce profits, although frequently terms such as 'financial sense' (Robichaud and Anantatmula 2011: 49) and 'competitive advantage' (Phua 2018: 417) are used as euphemisms for profitmaking. One rare exception to this discourse is Clarke et al.'s (2017) critique of how what amounts to market managerialism has undermined Vocational Education and Training (VET) in sustainable construction. Drawing on the example of the UK, they explain how senior managers have used their top-down authority to reduce labour costs through labour-only contracting and extensive sub-contracting (Clarke et al. 2017).

These practices undermine environmental sustainability by reducing training investment, instilling occupational disconnects at crucial component interfaces where energy is lost (and so more carbon emitted), and reducing capacities to ensure building quality (Clarke et al. 2017: 82–5). Clarke et al. (2017) conclude by proposing a need for 'radical transformation of both the structure of the industry and the VET system, to be achieved by enhancing the role of labour as an agent of production' (88). Such calls are laudable, but what might a radically different, less managerial, less market-based industry actually look like? And can such an industry help develop sustainable construction solutions? Taken together, all three cases I have presented offer some suggestions. The Ulma Group construction cooperatives are certainly not carbon-neutral firms (particularly given their focus on concrete products), but through the Mondragón Corporation, these firms do invest significantly in renewable community energy schemes, zero-carbon technology research and development, and sustainable construction education. Both the Nant-y-Cwm Steiner School (Kraftl 2006a, 2006b) and Colin Ward's (1976, 1996, 2004) anarchist housing studies also demonstrate how construction can take place in ways that significantly reduce ecological impacts.

These three topics are merely illustrative of the myriad ways alternative construction organizing can open up a new research agenda for construction organization. What is required is further consideration of what alternatives are possible, what they might accomplish, and how we can bring them into being. As Colin Ward might say, the snow of market managerialism has meant that such alternatives have remained hidden, but the seed has always existed.

References

Al-Bayati, A., Abudayyeh, O., Fredericks, T., and Butt, S. (2017) Managing cultural diversity at U.S. construction sites: Hispanic workers' perspectives, *Journal of Construction Engineering Management*, 143(9). https:// doi .org/ 10 .1061/ (asce)co .1943-7862.0001359

Berg, J., Thuesen, C., Ernsten, S., and Jensen, P. (2021) Reconfiguring the construction value chain: Analysing key sources of friction in the business model archetypes of AEC companies in strategic partnerships, *Construction Management and Economics*, 39(6): 533–48.

Bousalham, Y., and Vidaillet, B. (2018) Contradiction, circumvention and instrumentalization of noble values: How competition undermines the potential of alternatives, *Organization*, 25(3): 401–27.

Bower, R. (2017) Forgotten plotlanders: Learning from the survival of lost informal housing in the UK, *Housing, Theory and Society*, 34(1): 79–105.

Bridges, D., Wulff, E., Bamberry, L., Krivokapic-Skoko, B., and Jenkins, S. (2020) Negotiating gender in the male-dominated skilled trades: A systematic literature review, *Construction Management and Economics*, 38(10): 894–916.

Burham, J. (1941/1972) *The Managerial Revolution: What Is Happening In the World*, John Day Company Inc: New York.

Cheney, J., Cruz, I., Peredo, A., and Nazareno, E. (2014) Worker cooperatives as an organizational alternative: Challenges, achievements and promise in business governance and ownership, *Organization*, 21(5): 591–603.

Clarke, L. (1992) *Building Capitalism: Historical Change and the Labour Process in the Production of the Built Environment*, Routledge: London.

Clarke, L., Gleeson, C., and Winch, C. (2017) What kind of expertise is needed for low energy construction?, *Construction Management and Economics*, 35(3): 78–89.

Construction Enquirer. (2018) Average salaries for construction managers top £80,000, *Construction Enquirer*. https://www.constructionenquirer.com/2018/03/12/average-salaries-for-construction-managers-top-80000/

Contu, A. (2019) Conflict and organization studies, *Organization Studies*, 40(10): 1445–62.

Dainty, A., Bagihole, B., and Neale, R. (2000) A grounded theory of women's career underachievement in large UK construction companies, *Construction Management and Economics*, 18(2): 239–50.

Del Fa, S., and Vasquez, C. (2019) Existing through differantiation: A Derridean approach to alternative organizations, M@n@gement, 22(4): 559–83.

Dey, P., Schneider, H., and Maier, F. (2016) Intermediary organisations and the hegemonisation of social entrepreneurship: fantasmatic articulations, constitutive quiescences, and moments of indeterminacy, *Organization Studies*, 37(10) 1451–72.

Duong, L., Wang, J., Wood, L., Reiners, T., and Koushan, M. (2021) The value of incremental environmental sustainability innovation in the construction industry: An event study, *Construction Management and Economics*, 39(5): 398–418.

Farias, C. (2017) Money is the root of all evil – Or is it? Recreating culture through everyday neutralizing practices, *Organization Studies*, 38(6): 775–93.

Flech, R., and Ngai, P. (2014) The challenge for Mondragón: Searching for the cooperative values in times of internationalization, *Organization*, 21(5): 666–82.

Francis, V. (2017) What influences professional women's career advancement in construction?, *Construction Management and Economics*, 35(5): 254–75.

Frenkel, M. (2008) The multinational corporation as a third space: Rethinking international management discourse on knowledge transfer through Homi Bhabha, *Academy of Management Review*, 33(4): 924–42.

Grubbauer, M. (2018) Building home futures: Materialities of construction and meanings of home in self-help building practices, In D. Sage and C. Vitry (Eds), *Societies under Construction: Geographies, Sociologies and Histories of Building*, Palgrave Macmillan: London, pp. 185–204.

Guardian (2010, 22 February) Colin Ward obituary, *The Guardian*. https:// www.theguardian.com/society/2010/feb/22/colin-ward-obituary

Hasan, A., Ghosh, A., Mahmood, M., and Thaheem, M. (2021) Scientometric review of the twenty-first century research on women in construction, *Journal of Management in Engineering*, 37(3). https://doi.org/10.1061/(ASCE)ME.1943-5479.0000887

Heras-Saizarbitoria, I. (2014) The ties that bind? Exploring the basic principles of worker-owned organizations in practice, *Organization*, 21(5): 645–65.

Hill, R., and Bowen, P. (1997) Sustainable construction: Principles and a framework for attainment, *Construction Management and Economics*, 15(3): 223–39.

Jameson, F. (2003) Future city, *New Left Review*, https://newleftreview.org/issues/ii21/articles/fredric-jameson-future-city
Jones, T., Shan, Y., & Goodrum, P. (2010) An investigation of corporate approaches to sustainability in the US engineering and construction industry, *Construction Management and Economics*, 28(9): 971–83.
Just, N., De Cock, C., and Schaefer, S. (2021) From antagonists to allies? Exploring the critical performativity of alternative organization, *Culture and Organization*, 27(2): 89–97.
Klikauer, T. (2013) *Managerialism*, Palgrave Macmillan: London.
Kraftl, P. (2006a) Building an idea: The material construction of an ideal childhood, *Transactions of the Institute of British Geographers*, 31(4): 488–504.
Kraftl, P. (2006b) Ecological architecture as performed art: Nant-y-Cwm Steiner School, Pembrokeshire, *Social and Cultural Geography*, 7(6): 927–48.
Kuitert, L., Volker, L., and Hermans, M. (2019) Taking on a wider view: Public value interests of construction clients in a changing construction industry, *Construction Management and Economics*, 37(5): 257–77.
Leiringer, R., Green, S., and Raja, J. (2009) Living up to the value agenda: The empirical realities of through-life value creation in construction, *Construction Management and Economics*, 27(3): 271–85.
Linear Recruitment. (2019) *Highest paying construction jobs in the UK for 2019*. https://www.linearrecruitment.co.uk/news/highest-paying-construction-jobs-uk
Loosemore, M., and Heng, B. (2017) Linking corporate social responsibility and organizational performance in the construction industry, *Construction Management and Economics*, 35(3): 90–105.
Löwstedt, M., and Sandberg, R. (2020) Standardizing the free and independent professional: The case of construction site managers in Sweden, *Engineering, Construction and Architectural Management*, 27(6): 1337–55.
Marcuse, H. (1964/1999) *One-Dimensional Man*, Routledge: London.
Mondragón. (2021a) Mondragón Annual Report 2021 (extended). https:// www .mondragon -corporation .com/ urtekotxostena/ dist/ docs/ eng/ annual -report -2021 -extended.pdf
Mondragón. (2021b) *What we do – Mondragón*. https://www.mondragon-corporation .com/en/we-do/
Ong, H., Chen, W., and Zainon, N. (2019) Developing a quality-embedded EVM tool to facilitate the iron triangle in architectural, construction, and engineering practices, *Journal of Construction Engineering Management*, 144(9). https://doi.org/10.1061/(ASCE)CO.1943-7862.0001533
Oxfam. (2021, 20 January) World's billionaires have more wealth than 4.6 billion people. https:// www .oxfam .org/ en/ press -releases/ worlds -billionaires -have -more -wealth-46-billion-people
Parker, M. (2002) *Against Management*, Polity Press: Cambridge.
Parker, M. (2018) *Shut Down the Business School: What's Wrong with Management Education*, Pluto Press: London.
Parker, M., Cheney, G., Fournier, V., and Land, C. (Eds) (2014a) *The Routledge Companion to Alternative Organization*, Routledge: London.
Parker, M., Cheney, G., Fournier, V., and Land, C. (2014b) Advanced capitalism: Its promise and failings, In M. Parker, G. Cheney, G. V. Fournier, and C. Land (Eds), *The Routledge Companion to Alternative Organization*, Routledge: London, pp. 3–17.
Parker, M., Cheney, G., Fournier, V., and Land, C. (2014c) The question of organization: A manifesto for Alternatives, *Ephemera*, 14(4): 623–38.

Parker, M., Fournier, V., and Reedy, P. (2007) *The Dictionary of Alternatives*, Zed Books: London.
Parker, S., and Parker, M. (2017) Antagonism, accommodation and agonism in Critical Management Studies: Alternative organizations as allies, *Human Relations*, 70(11): 1366–87.
Phua, F. (2018) The role of organizational climate in socially embedding construction firms' sustainability goals, *Construction Management and Economics*, 36(7): 409–21.
Priola, V., Lasio, D., De Simone, S., and Serri, F. (2014) The sound of silence. Lesbian, gay, bisexual and transgender discrimination in 'inclusive organizations', *British Journal of Management*, 25: 488–502.
Reinecke, J. (2018) Social movements and prefigurative organizing: Confronting entrenched inequalities in Occupy London, *Organization Studies*, 39(9): 1299–1321.
Robichaud, L., and Anantatmula, V. (2011) Greening project management practices for sustainable construction, *Journal of Management in Engineering*, 27(1). https://doi .org/10.1061/(ASCE)ME.1943-5479.0000030
Rodgers, D., Peterson, J., and Sanderson, J. (2016) Commemorating alternative organizations and marginalized spaces: The case of forgotten Finntowns, *Organization*, 23(1): 90–113.
Russell, A., Tran, N., and Staub-French, S. (2014) Searching for value: Construction strategy exploration and linear planning, *Construction Management and Economics*, 32(6): 520–47.
Sage, D., and Vitry, C. (2018) Introduction: Societies under Construction, In D. Sage and C. Vitry (Eds), *Societies under Construction: Geographies, Sociologies and Histories of Building*, Palgrave Macmillan: London, pp. 1–50.
Sage, D., Vitry, C., and Dainty, A. (2019) Exploring the organizational proliferation of new technologies: An actor-network theory, *Organization Studies*, 41(3): 345–63.
Sang, K., and Powell, A. (2012) Equality, diversity, inclusion and work–life balance in construction, In A. Dainty and M. Loosemore (Eds), *Human Resource Management in Construction: Critical Perspectives*, Routledge, Abingdon, pp. 163–96.
Steiner. (2021) What is Steiner education? https:// www .steinerwaldorf .org/ steiner -education/what-is-steiner-education/
Stevenson, H., and Wood, P. (2013) Markets, managerialism and teachers' work: the invisible hand of high stakes testing in England, *International Education Journal: Comparative Perspectives*, 12(2): 42–61.
Styhre, A. (2011) In the circuit of credibility: Construction workers and the norms of 'a good job', *Construction Management and Economics*, 29(2): 199–209.
Tan, Y., Ochoa, J., Langston, C., and Shen, L. (2015) An empirical study on the relationship between sustainability and performance and business competitiveness of international construction contractors, *Journal of Cleaner Production*, 93: 273–8.
Tang, P., Cass, D., and Mukherjee, A. (2013) Investigating the effect of construction management strategies on project greenhouse gas emissions using interactive simulation, *Journal of Cleaner Production*, 54: 78–88.
Ulma Construction. (2021) *Engineering and applications – Ulma Construction, Ulma Group*. https:// www .ulmaconstruction .com/ en/ services/ engineering -and -applications
Ulma Architectural Solutions. (2021) *Ulma Architectural Solutions, Ulma Group*. https://www.ulmaarchitectural.com/en-gb
Uzuriaga, A., Freundlich, F., and Gago, M. (2018) Ulma Architectural Solutions: A case from the Mondragon Cooperative Group, In D. Berry and T. Kato (Eds),

Employee Ownership and Employee Involvement at Work, Emerald Publishing, Bingley, pp. 37–76.
Vitry, C. (2020) Queering space and organizing with Sara Ahmed's Queer Phenomenology, *Gender, Work and Organization*, 28(3): 935–49.
Ward, C. (1966) *Anarchism as a theory of organization*, The Anarchist Library. http://theanarchistlibrary.org/library/colin-ward-anarchism-as-a-theory-of-organization
Ward, C. (1973) *Anarchism In Action*, Freedom Press: London.
Ward, C. (1976) *Housing: An Anarchist Approach*, Freedom Press: London.
Ward, C. (1996) *Talking to Architects*, Freedom Press: London.
Ward, C. (2004) *The hidden history of housing*, The Anarchist Library. https://theanarchistlibrary.org/library/colin-ward-the-hidden-history-of-housing
Watts, J. (2007). Porn, pride and pessimism: Experiences of women working in professional construction roles. *Work, Employment & Society*, 21(2): 299–316.
Wright, T. (2013) Uncovering sexuality and gender: An intersectional examination of women's experience in UK construction, *Construction Management and Economics*, 31(8): 832–44.
Zanoni, P. (2020) Prefiguring alternatives through the articulation of post- and anti-capitalistic politics: An introduction to three additional papers and a reflection, *Organization*, 27(1): 3–16.

Index